Marine functional food

Marine functional food

edited by: J.B. Luten

Wageningen Academic
Publishers

ISBN 978-90-8686-078-4

Cover design:
Oddvar Dahl, Nofima, Norway

Cover photos:
IKO/ScanStockPhoto,
T.J.Udvang/ScanStockPhoto
and Frank Gregersen, Nofima, Norway

First published, 2009

Wageningen Academic Publishers
The Netherlands, 2009

Foreword

Stories, myths, and sayings are a universal characteristic of Nordic cultures. Passed on verbally through time from one generation to the next, the validity of these proverbial expressions has often been questioned – some concluded to be facts, others to be misconceptions. "Fish is good for you" *was* merely a widely known saying. Today it is a fact proved by research.

In the Nordic region, marine-based food has throughout history played an important, and sometimes lifesaving, role in peoples' diets, as well as functioning as a significant contributor to the Nordic economy. Today, our knowledge concerning the relationship between food, lifestyle and health is increasing, hence raised consumer awareness about how food contributes to ones health. Consequently, this also leads to new market opportunities, which together with emerging technologies, developments and globalization require new thinking and thus innovation.

Over the last fifteen years, the Nordic Innovation Centre has launched several programs, the main goal of which has been to enhance innovation in the Nordic food industry and increase its competitiveness. Previous collaborative arrangements between the industry, R&D organisations and universities have provided valuable results. Nevertheless, one has also noticed the emerging trend of food and functionality. At this present moment, functional food is one of the fastest growing markets within the food industry. However, there are challenges to be faced when entering this market, such as documentation of health claims, communicating functionality and achieving consumer credibility. Several marine-based food components have gained interest as potentially bioactive and health promoting ingredients. Still, substantial evidence for health beneficial claims is limited. As an investor and co-operative participant in the project *The Nordic Network for Marine Functional Food* (MARIFUNC), the Nordic Innovation Centre is happy to see that the Nordic network has been established and we are certain that this will have a strengthening effect upon the marine-based food industries in the Nordic countries.

This book portrays how raw material used in new contexts can create a more generic consumer audience. It also nicely illustrates how knowledge can contribute to innovations by establishing a foundation for the development of new products, thus creating new market opportunities and value for stakeholders.

I hope you will enjoy reading this book as much as I did.

Yours sincerely,

Ivar H. Kristensen
Managing director
Nordic Innovation Centre, Oslo, Norway

Table of contents

Preface

This book is based on the outcome of the activities of the Nordic Network on Marine Functional Food (MARIFUNC). MARIFUNC is one of the six projects funded by the Nordic Innovation Centre (NICe) in the area of functional foods. It was launched on 1st June 2006. The project duration was 2.5 years. MARIFUNC was a consortium with 13 partners including research institutes, universities and private companies in 5 Nordic countries (Norway, Sweden, Denmark, Finland and Iceland).

Scientific and technological developments in the field of food have led to a marked shift in the way consumers perceive food and health. There is a growing awareness that the dietary source and form of food may affect the overall health of the consumer. The role of food as an agent for improving health has initiated the development of new classes of food: functional foods. The main activity in MARIFUNC was to create an overview of the role fish- and seafood-derived components play in relation to health, the quality of the functional seafood components, the processing of functional seafood components and the consumer's attitude to (marine) functional foods. The contributions from the MARIFUNC project in this book are the results of the combined efforts of more than 20 specialists drawn from the different partners in the consortium. One contribution in the book is taken from another functional food NICe project called 'SMEs commercializing healthy nutrition'. With this book we aim to reach a broad group of readers, from students and scientists to research managers in the (sea)food industry.

I wish to express my thanks to all the authors who contributed to the chapters of this book. Of those, associate Prof Dr. Ingrid Undeland deserves a special thanks for her committed work in bringing together the various and complex issues of seafood and health. Experts have refereed all chapters. I would like to thank the reviewers for giving their valuable comments on how to improve the content of each chapter. Thanks also to Sigridur Thormodsdottir and Hilde Helgesen, senior innovation advisors at NICe for their interest in the MARIFUNC project. Last but not least, thanks go to my Nofima colleague Heidi K. Trige, for her secretarial help with all the manuscripts and Oddvar Dahl for his creativity in designing the cover of the book.

Joop Luten
Coordinator MARIFUNC
Nofima Marine, Tromsø, Norway

Seafood and health: what is the full story?

Ingrid Undeland[1], Helen Lindqvist[1], Yun Chen-Yun[2], Eva Falch[3], Alfonso Ramel[4], Marie Cooper[5], Asbjørn Gildberg[5], Joop Luten[5], Even Stenberg[5], Henrik Hauch Nielsen[6] and Edel Elvevoll[7]

[1]*Chalmers University of Technology, Food Science, 41296 Göteborg, Sweden*
[2]*Sahlgrens University Hospital, Department of Clinical Physiology, 1345 Göteborg, Sweden*
[3]*SINTEF Fisheries and Aquaculture, 7465 Trondheim, Norway*
[4]*University of Island, Landspitali University Hospital, 01 Reykjavik, Island.*
[5]*Nofima Marine, P.O. Box 6122, 9191 Tromsø, Norway*
[6]*National Institute of Aquatic Resources, DTU, 2800 Kgs.Lyngby, Denmark*
[7]*Norwegian College of Fishery Science, Institute of Marine Biotechnology, University of Tromsø, Norway*

1. General introduction

Although the beneficial effects of seafood have been well known since the 1950´s, there has been an overwhelming focus on the long chain n-3 polyunsaturated fatty acids (LC n-3 PUFA), eicosapentaenoic acid (EPA) and docosapentaenoic acid (DHA) as the only carriers of the documented health effects. As a result, the words 'omega-3' and 'fish' are nowadays very often used as synonyms, even in the scientific literature. Without ignoring the multitude of studies that have indeed shown strongly positive effects from LC n-3 PUFA on various diseases/risk factors, we wish to stress in this chapter that consumption of seafood also ensures several important nutrients and micronutrients beyond the LC n-3 PUFA. On a wet weight basis, these 'non-n-3 compounds' usually contribute to 95-99.5% of the edible parts of seafood. In the past 20 years, a number of scientists have speculated on the likelihood that such non-n-3 compounds also contributes to the cardioprotection and neuroprotection documented e.g. from a low to moderate fish consumption (Kromhout *et al.* 1985; Marckmann and Gronbaek 1999; Elvevoll and Österud 2003; Ouellet *et al.* 2007; Thorsdottir *et al.* 2007; Tremblay *et al.* 2007). Also, a recent review (Hooper *et al.* 2006) concluded that LC n-3 PUFA alone do not have a clear effect on total mortality, combined cardiovascular events, or cancer. However, positive effects on these end-points were seen when only taking into account the fish-based studies they reviewed. It has been discussed that the latter was due to a heavy impact from the so-called DART II study (Burr *et al.* 2003) where LC n-3 PUFA intake did not lead to reduced cardiovascular mortality in humans.

Despite the listed indications that non-n-3 compounds from seafoods might affect certain diseases in a positive way, it is obvious that, when screening the literature, clear evidence for such effects is lacking. Most indications are from simplistic *in vitro* models, or in the best case, from animal studies which are hard to extrapolate to humans. With this background, the first section of this chapter has primarily been prepared to create an overview of available evidence for possible health benefits of whole seafood (mainly fish muscle) and non-n-3 seafood derived compounds. Among the latter, we have focused on vitamin D, proteins, peptides, amino acids and selenium. However, to give a holistic view of the tentative contribution from fish consumption on human health, we have also included the latest news about effects from the LC n-3 PUFA and also about marine phospholipids as alternative carriers of LC n-3 PUFA.

In a second section of this chapter, the potential health effects that can be achieved by utilising by-products of fish/shellfish are discussed together with health effects achieved when fortifying non-fish products with fish-derived components. In the former part we discuss chitin, chitosan, glucosamine and chondroitin sulphate. In the latter part, we have chosen LC n-3 PUFA fortification since this is by far the most developed area. The chapter is ended by a section dealing with potential risks associated with fish/fish component intake, as this question is automatically raised when health benefits of fish are discussed.

We also wish to stress that some of the compounds reviewed in this chapter are *generally* considered as healthy, regardless of the source they are extracted from. Information regarding their health effects is therefore not taken exclusively from fish-based studies. Examples of this category are vitamin D, selenium (Se) and chondroitin sulphate. Other compounds dealt with here are more specific to seafood, and are not commonly found elsewhere; e.g. the LC n-3 PUFA and the seafood proteins.

We have used a broad approach in the sense that a wide span of effects has been considered, ranging from those linked to cardiovascular diseases (CVD) and other metabolic diseases, to those related to inflammatory diseases, brain functions (well-being) and bone health. We have made an attempt to highlight from where the scientific evidence has emerged, i.e. from human intervention studies, from human observational studies or from so-called supportive studies (e.g. animal studies, *in vitro* cell and molecular studies, genotype studies and mechanism modelling studies). Intervention studies include the randomised controlled trial (RCT) in healthy subjects, clinical trials that are made in patients and physiological as well as psychological trials. The latter include healthy subjects and provide means for a detailed characterisation of effects and their possible mechanistic bases. A highly controlled environment must be used to allow integration of cellular and molecular studies into whole body metabolism. Among all these studies, the RCT is thought to provide the best standard of evidence (Aggett *et al.* 2005). In these studies people are allocated at random to receive one of two or more interventions, one of which would usually be a control. Observational studies, which are often referred to as epidemiological studies, include prospective (cohort) studies, case control, cross-sectional (analytical) and ecological studies. If all of these study types are performed equally well, findings from prospective cohort studies should receive more weight than data from other observational studies. Animal studies may provide insights that can be used to design human studies. They can also be deployed where the use of human subjects is unethical. Animal studies may provide supporting evidence in cases where the comparability of certain parameters between animal and human studies has been established. *In vitro* cellular and molecular studies often provide supportive evidence of the effect of food/food components on cell function. Such studies cannot on their own indicate a health benefit that might be the basis for a health claim, for instance. However, *in vitro* studies can provide important mechanistic information that can lead to the identification of markers to use in other studies. A main difference between knowledge on LC n-3 PUFA and other seafood-derived compounds lies in the kinds of studies that have so far been used to investigate their effects. While a multitude of human intervention studies have been performed on the LC-n-3 PUFA, such studies are very scarce for the other compounds.

Since the present chapter is meant to give a comprised overview regarding seafood and health, our starting point in searching the literature was to collect reviews from the last five years. In the fields of whole fish, LC n-3 PUFA and vitamin D this strategy resulted in a good basis for writing, at least in the areas of CVD. However, in other fields, where few or no recent reviews existed, older reviews and original papers were used and cited. Within some of the fields addressed here (e.g. seafood proteins and seafood phospholipids), the nutritional literature is very limited. This is why no firm criteria were used to make a selection from the available papers. Our aim was rather to highlight the few attempts that have been made to study non-n-3 components in relation to human health. The main databases used in searching the literature have been PubMed, Science Direct, ISI Web of Science and Google Scholar.

2. Consumption of whole fish muscle

2.1 CVD[1] and metabolic syndrome related diseases

2.1.1 CVD mortality

Several recent reviews arrive at the conclusion that fish consumption decreases the risk of CHD[2] and CHD mortality (Burr 1993; Undeland *et al.* 2004; König *et al.* 2005;

[1] Cardiovascular disease.
[2] Coronary heart disease.

Psota *et al.* 2006) in primary prevention. Two of them evaluated cohorts[3] and came up with similar conclusions being that low consumption (1-3 times/month) of fish reduces the relative risk (RR) of CHD mortality by about 11-17% compared to no fish consumption. Increasing fish consumption will further decrease the risk of CHD mortality (Burr 1993; Psota *et al.* 2006). Wang *et al.* (2006) concluded that increased consumption of fish reduces the rates of all causes of mortality[4], sudden cardiac death and possible stroke in both primary and secondary prevention. The evidence is stronger in secondary prevention (Whelton *et al.* 2004).

2.1.2 Stroke[5]

Two reviews concluded that 1-3 fish servings /month decrease the RR of stroke by 12-13% and with higher intake, the RR further decreases slightly. Ischemic[6] stroke is also reduced by fish intake (He *et al.* 2004a; Bouzan *et al.* 2005). Ischemic stroke, involves the blocking of arteries leading to the brain caused by atherosclerosis and thrombosis. The mechanisms in preventing ischemic stroke are thus thought to be the same as for CHD, including reduction of platelet aggregation, decreased fibrinogen levels and improved endothelial function from LC n-3 PUFA.

2.1.3 Arrhythmia[7]

An observational study concluded that intake of tuna or other fish (broiled or baked), but not fried fish, correlated positively with LC n-3 PUFA of the plasma PL and also with decreased incidence of atrial fibrillation[8] (Mozaffarian *et al.* 2006). Other observational studies have shown that fish consumption is associated with decreased heart rate in men (Dallongeville *et al.* 2003; Mozaffarian *et al.* 2006) and that high fish consumption compared to no intake decrease the risk of arrhythmias (Mozaffarin *et al.* 2006; Chrysohoou *et al.* 2007).

2.1.4 Blood pressure

The hypotensive effect of fish seems to be of minor importance, even if some studies have reported this effect (Bao *et al.* 1998; Mori *et al.* 2004; Undeland *et al.* 2004). Three recent intervention studies on fish intake showed reduced blood pressure with a daily fish meal (Lara *et al.* 2006). The main active compounds behind reduced blood pressure are believed to be the LC n-3 PUFA. However, studies on fish

[3] Correlational research study that involves repeated observations of the same items over long periods of time - often many decades.
[4] Mortality rate is a measure of the number of deaths in a given population.
[5] Rapidly developing loss of brain function due to an interruption in the blood supply to all or part of the brain.
[6] Lack of oxygen in the tissue.
[7] The electrical activity of the heart is irregular or is faster or slower than normal.
[8] Abnormal heart rhythm which involves the two upper chambers of the heart.

oil have commonly investigated the intake of high doses of EPA and DHA, which are unlikely to be ingested by eating fish. This might explain the relatively minor effect of fish intake on blood pressure compared to that given by fish oil. Also fish proteins (mainly cod) have been found to decrease blood pressure in animal studies by affecting the angiotensin converting enzyme, ACE (Ait *et al.* 2003, 2005).

2.1.5 Atherosclerosis[9]

In a study from Japan, carotid Intima-Media Thickness (IMT)[10] was found to be larger among people from a farming village than a fishing village. IMT seemed to be related to the amount of LC n-3 PUFA in the plasma (Yamada *et al.* 2000). This relationship was also seen in a study where postmenopausal women eating fish >2 times/week or fatty fish >1 time/week were compared with women who ate less fish. The high fish consuming women had less stenosis, a smaller decrease of the coronary artery diameter and less new lesions than the low fish-consuming women (Erkkila *et al.* 2004).

2.1.6 Blood lipids

Blood triacylglycerol (TAG) has been shown to decrease with fish intake in many studies. It is believed that this effect is also mainly mediated by the LC n-3 PUFA of the fish (Undeland *et al.* 2004). A reason that the effect has not shown up in all fish-based studies (Lindqvist *et al.* 2007; Lindqvist *et al.* 2008) could therefore be that these studies have comprised a low intake of fatty fish or intake of lean fish. High-density lipoprotein (HDL)[11], or more specifically HDL_2, has been shown to increase with fish intake (Undeland *et al.* 2004). The TAG lowering and HDL-increasing effects from fatty fish have been confirmed in several recent intervention studies (Lara *et al.* 2006). The effect on low density lipoprotein (LDL)[12] is uncertain since studies have shown both decreased (Beauchesne-Rondeau *et al.* 2003; Lara *et al.* 2006), unaffected (Seierstad *et al.* 2005) and increased plasma LDL levels (Mori *et al.* 1999a) following fish intake. Few studies on fish intake and blood lipids have focused on lean fish.

2.1.7 Haemostasis[13]

It is well known that high intake of fatty fish rich in EPA and DHA increases bleeding time (Knapp 1997). The mechanism involves the arachidonic acid pathway and includes a decreased risk for blood clotting (thrombosis) that could cause stroke

[9] Atherosclerosis is a disease where the formation of multiple plaques affects our arterial blood vessels.
[10] The thickness of a blood vessel from the heart, used as a measurement for atherosclerosis progression.
[11] Simplified: transports cholesterol away from the blood vessels.
[12] Simplified: transports cholestreol to the blood vessels.
[13] Hemostasis can refer to the physiological process whereby bleeding is halted, thus protecting the integrity of the vascular system after tissue injury.

or heart failure (Thorngren 1983; Undeland *et al.* 2004). There are no consistent findings on haemostatic or fibrinolytic factors from fish intake, but in a recent paper, Din *et al.* (2008) demonstrated that 500 g mackerel/week for four weeks reduced platelet aggregation in man. The study used flow cytometry to measure platelet aggregation which is considered to be a more reliable method than the previously used methods. There was an inverse relationship between plasma PL LC n-3 PUFA and platelet-monocyte aggregation (Din *et al.* 2008).

2.1.8 Inflammation

One observational study compared a group that reported '>300g fish/week' with a group reporting 'no fish consumption' and found that the group with high fish intake had significantly decreased inflammation markers such as C-reactive protein (CRP), interleukin-6 (IL-6), tumor necrosis factor-α (TNF-α), serum amyloid A (SAA) and white blood cell count (Zampelas *et al.* 2005). In haemodialysis[14] patients, serum CRP was only decreased after sardine supplementation in the tertile with the highest CRP levels (Moreira *et al.* 2007). In a salmon intervention study, no effects from the intake of salmon low in EPA/DHA on cell adhesion molecules such as E-selectin, P-selectin, intercellular adhesion molecule-1 (ICAM-1) and the inflammation marker CRP were shown. However, with salmon rich in EPA/DHA, vascular adhesion molecule-1 (VCAM-1) and IL-6 was decreased. TNF-α decreased with both low and high EPA/DHA-containing salmon (Seierstad *et al.* 2005).

2.1.9 Oxidative stress

A few intervention studies focused on fish have tried to estimate oxidative stress by measuring oxidative products like urinary F2-isoprostanes, urine or plasma TBARS and plasma ox-LDL (Nelson *et al.* 1993; Mori *et al.* 1999b; Seierstad *et al.* 2005). No conclusions can be drawn from these few studies as to whether fish in the diet increases or decreases oxidation product formation *in vivo*.

2.1.10 Obesity, weight loss and appetite regulation

Two intervention studies on the effect of fish intake on leptin[15], glucose and insulin showed no effects, but a positive trend was shown for adiponectin[16] (Mori *et al.* 1999a, 2004). Fish intake in combination with weight loss produced a greater effect on leptin than weight loss alone. No differences were shown after fish or beef intake on appetite as rated by Visual Analogue Scale. However, subjects displayed an 11% reduction in energy intake at the subsequent meal after the fish meal (Borzoei *et al.*

[14] Hemodialysis is a method for removing waste products such as potassium and urea, as well as free water from the blood when the kidneys are in renal failure.
[15] Hormone that regulates energy intake and energy expenditure.
[16] Modulates a number of metabolic processes, including glucose regulation and fatty acid catabolism.

2006). A recent study showed greater weight loss in young overweight men when including lean fish, fatty fish or fish oil in an energy-restricted diet compared to an isocaloric control diet (Thorsdottir *et al.* 2007). Another recent study (Ramel *et al.* 2008), showed a dose-response relationship between cod consumption and weight loss in an intervention study of 126 subjects (20-40 years, BMI 27.5-32.5). The subjects were divided into 3 groups consuming from 0g cod/wk to 150 g cod 5 times/wk.

2.2 Cancer

2.2.1 Non-hormone related cancers

Prospective and case-control studies either do not show an association between fish intake and cancer risk, or show reduced risk at high fish intakes (De Deckere 1999). In 2002, Lund concluded that there were few studies and not of enough quality to draw any conclusions on the relation between fish intake and cancer (Lund 2002). The World Cancer Research Fund arrived at the same conclusion in 2007; there is limited evidence suggesting that fish consumption protects against colorectal cancer. The heterogenic results may be partially explained by the varying fish species and preparations of the fish used in the different studies. Components in fish that might prevent cancer are LC n-3 PUFA, selenium or vitamin D (World Cancer Research Fund 2007). A pooled analysis of thirteen case control studies showed no increased risk of thyroid cancer with fish consumption and that fish intake may instead have a positive effect in areas with iodine deficiency (Bosetti *et al.* 2001).

Observational studies on fish consumption have shown protection for risk of digestion tract cancer: oral cavity and pharynx[17], oesophagus[18], stomach, colon and rectum. Protection has also been shown for ovary, pancreas, larynx and endometrial[19] cancers. However, for multiple myeloma[20] and bladder cancer, only trends have been shown. No relation was shown for prostate, kidney, lymphomas, gallbladder, breast, liver and thyroid cancers (Fernandez *et al.* 1999). Several recent studies have confirmed that fish intake decreases the risk for colorectal cancer, and that red meat has the opposite effect (Norat *et al.* 2005; Siezen *et al.* 2006). However, two studies performed in Japan showed either no correlation or no significant inverse relationship between colon or rectal cancer with fish intake (Yang *et al.* 2003; Kimura *et al.* 2007). It has been suggested that the influence from fish intake on the risk of adenomas[21] may differ by genetic variation in the enzyme cyclo-oxygenase 1, COX-1 (Poole *et al.* 2007).

[17] Throat.
[18] Gullet.
[19] Uterus.
[20] Cancer of immune system cells in bone marrow.
[21] Collection of growths of glandular origin (non-cancerous but can become cancerous).

2.2.2 Hormone related cancers

Evidence from epidemiological and observational studies on hormone-dependent cancers remains unclear (Stripp *et al.* 2003; Terry *et al.* 2003; Engeset *et al.* 2006), but some population studies with high intake of fish during many years is associated with reduced risk for breast cancer (De Deckere 1999).

2.3 Inflammatory diseases

2.3.1 Rheumatoid arthritis[22]

There is evidence that the diet could influence the onset of rheumatoid arthritis and that consumption of fatty fish, just like the consumption e.g. of fruits and vegetables, could have a protective effect (Pattison *et al.* 2004). Fish oil has shown to have anti-inflammatory effects and positive effects on rheumatoid arthritis (Calder 2006), but no intervention studies on fish intake and rheumatoid arthritis have been found.

2.3.2 Allergy

One observational study concludes that regular fish consumption before age one appears to be associated with a reduced risk of allergic disease, sensitisation to food and inhalant allergens during the first four years of life. Sensation to fish is rare, even when fish is introduced early (Kull *et al.* 2006). Several very recent studies have shown a relationship between maternal fish consumption during pregnancy and decreased risk for early childhood atopy-related outcomes such as asthma, allergic sensitisations for food, dust and eczema (Salam *et al.* 2005; Calvani *et al.* 2006; Romieu *et al.* 2007; Sausenthaler *et al.* 2007; Willers *et al.* 2007).

2.4 Brain

2.4.1 Cognitive decline/dementia

Two reviews, unfortunately based on very few studies, indicate that fish consumption is associated with less cognitive decline or dementia (Friedland 2003; Kalmijn 2000). In populations 65 years and older, several observational studies have found a relationship between cognitive decline or development of Alzheimer's disease and low fish intake (Morris *et al.* 2003, 2005; Kalmijn *et al.* 2004; Huang *et al.* 2005). In one of these studies lean fried fish had no effect, but fatty fish >2 times/week compared to fish <1 time/ month decreased the risk for dementia by 28% (Huang *et al.* 2005). Barberger-Gateau *et al.* (2005) remarked that socioeconomic status,

[22] Chronic, inflammatory autoimmune disorder that causes the immune system to attack the joints.

dietary habits, depression and vascular risk factors could act as confounders[23] in the relationship between fish consumption and risk of dementia. Some, but not all of these confounders, have been taken into account in the studies on cognitive decline.

2.4.2 Mood/depression

A few observational studies have concluded that fish intake is associated with a lower incidence of depression or lower self-reported mental health status (Tanskanen *et al.* 2001; Silvers and Scott 2002; Timonen *et al.* 2004). Two out of these studies found that females, but not males, had a significantly higher risk of developing depression when comparing infrequent fish eaters with regular fish eaters. Two studies on postnatal[24] depression showed no effect of fish intake. However, in one of these studies, there was a very low intake of fatty fish among the fish eaters. In the other one, a trend towards a positive effect from fish consumption was seen (Browne *et al.* 2006; Miyake *et al.* 2006).

2.4.3 Cognitive development

Two observational studies have shown that high maternal fish intake during pregnancy and infant's fish intake during the first year is associated with modestly, but consistently, higher cognitive, fine motor, communication and social developmental scores (Daniels *et al.* 2004; Hibbeln *et al.* 2007).

2.4.4 Pregnancy

Only one observational study was found on fish intake and pregnancy duration. This study concluded that never consuming fish in the first two trimesters of pregnancy was a strong risk factor for preterm delivery (Olsen *et al.* 2006).

---

[23] Confounder is an extraneous variable in a statistical model that correlates (positively or negatively) with both the dependent variable and the independent variable.

[24] Postnatal is the period beginning immediately after the birth of a child and extending for about six weeks.

Table 1. *Overview of reviews, observational studies and human intervention studies on fish and human health cited in the present review.*

End point	Reviews	Observational studies	Interventions
CVD/metabolic syndrome related			
CVD mortality	Burr 1993; He *et al.* 2004b; Undeland *et al.* 2004 Whelton *et al.* 2004; Konig *et al.* 2005; Psota *et al.* 2006	Hunter *et al.* 1988; Burr *et al.* 1989; Zhang *et al.* 1999; Yuan *et al.* 2001; Hu *et al.* 2003; Nakamura *et al.* 2005	
Stroke	Skerret and Hennekens 2003; He *et al.* 2004a; Bouzan *et al.* 2005	Iso *et al.* 2001; He *et al.* 2002; Mozaffarian *et al.* 2005a; Myint *et al.* 2006	
CVD	Deslypere 1990; Nestel 2001; Kris-Etherton *et al.* 2003; Mozaffarian *et al.* 2003; Undeland *et al.* 2004; Whelton *et al.* 2004; Panagiotakos *et al.* 2005b; Mozaffarian and Rimm 2006; Psota *et al.* 2006; Wang *et al.* 2006	Burr and Fehily 1991; Burr 1992; Mozaffarin *et al.* 2003; Panagiotakos *et al.* 2005a; Iso *et al.* 2006	Lara *et al.* 2006
Arrhythmia		Mozaffarin *et al.* 2004; Dallongeville *et al.* 2003; Mozaffarin *et al.* 2006; Chrysohoou *et al.* 2007	
Blood pressure	Undeland *et al.* 2004		Bao *et al.* 1998; Ait Yahia *et al.* 2003; Mori *et al.* 2004; Ait Yahia *et al.* 2005; Lara *et al.* 2006
Atherosclerosis		Yamada *et al.* 2000; Erkkila *et al.* 2004	
Blood lipids	Undeland *et al.* 2004	Bulliyya 2000; Bulliyya 2002	Lacaille *et al.* 2000; Beauchesne-Rondeau *et al.* 2003; Li *et al.* 2004; Lindqvist *et al.* 2007; Lindqvist *et al.* 2008
Haemostasis	Thorngren 1983; Undeland *et al.* 2004		Muller *et al.* 1989; Dunstan *et al.* 1999; Din *et al.* 2008
Inflammation		Zampelas *et al.* 2005	Seierstad *et al.* 2005; Moreira *et al.* 2007

Table 1. *Continued.*

End point	Reviews	Observational studies	Interventions
Oxidative stress			Nelson *et al.* 1993; Mori *et al.* 1999b; Seierstad *et al.* 2005
Obesity/weight loss		Nkonfjock and Receveur 2003	Mori *et al.* 1999a; Mori *et al.* 2004; Borzoie *et al.* 2006; Thorsdottir *et al.* 2007; Ramel *et al.* 2008
Diabetes II		Hu *et al.* 2003; Nkondjock and Receveur 2003	Dunstan *et al.* 1997; Mori *et al.* 1999a
Appetite regulation/ satiety			Borzoei *et al.* 2006
Cancer			
Cancer general	de Deckere 1999; Lund 2002; Terry *et al.* 2003	Fernandez *et al.* 1999; Tavani *et al.* 2005; Siezen *et al.* 2006	
Digestive tract	de Deckere 1999; Bosetti *et al.* 2001; World Cancer Research Fund 2007	Yang *et al.* 2003; Kobayashi *et al.* 2004; Luchtenborg *et al.* 2005; Norat *et al.* 2005; Kimura *et al.* 2007; Poole *et al.* 2007	
Hormone related	de Deckere 1999; Terry *et al.* 2003	Stripp *et al.* 2003; Engeset *et al.* 2006	
Inflammatory diseases			
Arthritis		Pedersen *et al.* 2005	
Psoriasis		Collier *et al.* 1993; Collier and Payne 1996	
Allergy		Salam *et al.* 2005; Calvani *et al.* 2006; Kull *et al.* 2006; Romieu *et al.* 2007; Sausenthaler *et al.* 2007; Willers *et al.* 2007	

Table 1. *Continued.*

End point	Reviews	Observational studies	Interventions
Brain			
Mood/ Depression		Tanskanen *et al.* 2001; Silvers and Scott 2002; Timonen *et al.* 2004; Browne *et al.* 2006; Miyake *et al.* 2006	Ness *et al.* 2003
Cognitive decline/ Dementia	Friedland 2003	Morris *et al.* 2003, 2005; Huang *et al.* 2005; Kalmijn *et al.* 2004; Barberger-Gateau *et al.* 2005	
Cognitive development		Daniels *et al.* 2004; Hibbeln *et al.* 2007	
Pregnancy			
Pregnancy duration		Olsen *et al.* 2006	

3. Consumption of seafood-derived lipid soluble compounds

3.1 LC n-3 PUFA ('omega-3 fatty acids')

3.1.1 CVD

Consuming fish or fish oil containing LC n-3 PUFA is associated with reduced rates of all-cause mortality, cardiac and sudden death, and possibly stroke both in patients with CVD and in the general population (Breslow 2006). The evidence for the benefits of fish oil is stronger in secondary- than in primary-prevention settings (Psota *et al.* 2006). Randomised control trials in the context of secondary prevention also indicate that the consumption of LC n-3 PUFA is protective at doses <1 g/d (Breslow 2006). The therapeutic effect appears to be due to suppression of fatal arrhythmias (Breslow 2006; London *et al.* 2007). Whether LC n-3 PUFA treatment can lead to the stabilisation of vulnerable atherosclerotic plaques is still under debate (Breslow 2006; Hamer and Steptoe 2006), and the effects on the progression of atherosclerosis, haemostatic activity and vascular inflammation remain equivocal (Hamer and Steptoe 2006). Studies of coronary artery restenosis[25] rates suggest

[25] Restenosis literally means the *re*occurrence of *stenosis* which is an abnormal narrowing in a blood vessel or other tubular organ or structure.

only a possible trend that LC n-3 PUFA supplementation may be beneficial (Balk *et al.* 2006a). The few studies that assessed degree of carotid IMT failed to conclusively demonstrate an effect of LC n-3 PUFA (Balk *et al.* 2006a). Only small non-significant improvements in exercise capacity, diagnostic test related to cardiac atherosclerosis, are reported with fish oil supplementation (Balk *et al.* 2006a). There is little evidence for a major antithrombotic effect of practical doses of LC n-3 PUFA on coronary thrombosis (Kristensen *et al.* 2001).

3.1.2 The metabolic syndrome

Central obesity, high fasting concentration of blood glucose, high blood pressure, increased triglycerides and decreased HDL-cholesterol are the most important features of the metabolic syndrome. Daily doses of >3 g LC n-3 PUFA results in decreased plasma TAG (Breslow 2006; Balk *et al.* 2006a). The dose-dependent beneficial effect of LC n-3 PUFA on TAG is particularly evident among people with a high basal level of TAG (Balk *et al.* 2006b). LC n-3 PUFA consumption also modestly improves HDL-cholesterol and increases LDL cholesterol levels (Balk *et al.* 2006b). The evidence regarding the effect of LC n-3 PUFA on highly sensitive CRP is inconclusive (Balk *et al.* 2006b). Increased intakes or supplements of LC n-3 PUFA may improve defects in insulin signalling and prevent alterations in glucose homeostasis and the further development of type 2 diabetes (Carpentier *et al.* 2006). The effects of LC n-3 PUFAs on inflammation, platelet activation, endothelial[26] function, and blood pressure may contribute to decreasing the burden of the metabolic syndrome (Carpentier *et al.* 2006; Mori 2006).

3.1.3 Cancer

A large body of literature spanning numerous cohorts from many countries and with different demographic characteristics does not provide evidence to suggest a significant association between LC n-3 PUFA and cancer incidence (MacLean *et al.* 2006).

3.1.4 Inflammatory diseases

At sufficiently high intakes, LC n-3 PUFA decrease the production of inflammatory eicosanoids, cytokines[27], reactive oxygen species and the expression of adhesion molecules. Evidence of their clinical efficacy is reasonably strong in some settings (e.g. in rheumatoid arthritis) but is weak in others (e.g. in inflammatory bowel diseases and asthma) (Calder 2006).

[26] The endothelium is the thin layer of cells that line the interior surface of blood vessels, forming an interface between circulating blood in the lumen and the rest of the vessel wall.

[27] Cytokines are a category of signalling proteins and glycoproteins that, like hormones and neurotransmitters, are used extensively in cellular communication.

3.1.5 Brain

3.1.5.1 Dementia and macular degeneration

It is not possible to make firm recommendations regarding LC n-3 PUFA and the prevention of dementia and macular degeneration (Johnson and Schaefer 2006).

3.1.5.2 Schizophrenia

Supplementation of the LC n-3 PUFA, especially EPA, in addition to one's existing medication has been found to decrease symptoms in schizophrenic patients and to improve the clinical outcome in relation to schizophrenia (Mahadik *et al.* 2006; Lakhan and Vieira 2008). The LC n-3 PUFA are found to prevent and restore destructive oxidative neuropathology. According to the authors, a combination of antioxidants and LC n-3 PUFA, particularly in the early stages of illness, when the brain has a high degree of neuroplasticity, may potentially be even more effective for long-term improvement in the clinical outcome of schizophrenia (Mahadik *et al.* 2006).

3.1.5.3 Depressed mood

In a review from 2006 (Appleton *et al.* 2006) it is stated that the available evidence provides fairly little support for the use of LC n-3 PUFAs to improve depressed mood. However, a more recent review (Clayton *et al.* 2007), concludes that there is evidence that LC n-3 PUFA may be involved in the aetiology of depression in adults, although to date there are no large controlled studies on children or adolescents. In a small randomised controlled trial on children, it was however shown that LC n-3 PUFA supplementation provided a small beneficial effect on depression over the placebo (Clayton *et al.* 2007).

3.1.5.4 ADHD and other psychiatric illnesses

Four placebo-controlled trials showed uncertain benefit of LC n-3 PUFA on attention-deficit hyperactivity disorder (ADHD) (Clayton *et al.* 2007). Single placebo-controlled trials showed no benefit from LC n-3 PUFA on autism or bipolar disorders (Clayton *et al.* 2007). The latter is a category of mood disorder defined by the presence of one or more episodes of abnormally elevated mood, clinically referred to as mania.

3.1.6 Effects during pregnancy

Studies assessing the influence of LC n-3 PUFA during pregnancy or early postpartum period on duration of gestation, infant size at birth, pre-eclampsia[28] and depression have been reported (Jensen 2006). No clear consensus exists regarding the effects of LC n-3 PUFA on any of these outcomes. The available data suggest a modest effect of these fatty acids on increasing gestational duration (Jensen 2006; Szajewska *et al.* 2006). Higher visual acuity after LC n-3 PUFA supplementation is a consistent finding in infants born preterm (Jensen 2006; Cheatham *et al.* 2006). For infants born at term, the results are less consistent (Cheatham *et al.* 2006). Randomised clinical trials studying the effects of LC n-3 PUFA on cognitive development revealed mixed results in both preterm and term samples. The inconsistent results could be related to the use of inappropriate measures of cognition (Cheatham *et al.* 2006).

3.1.7 The n-6/n-3 ratio

In this section about LC n-3 PUFA, it must also be stressed that a point of controversy has been whether the absolute mass of LC n-3 PUFA consumed, or the n-6/n-3 ratio of the diet, should be the first consideration when contemplating lifelong dietary habits affecting cardiovascular benefit from their intake. Some scientists believe that the LC n-3 PUFA alone, especially EPA and DHA, appear to hold the greatest promise (Wijendran and Hayes 2004; Harris 2006). This is based on the observation that there are few human experimental and clinical trial data to support the importance of the n-6/n-3 ratio. Others stress that the large increase in the n-6/n-3 ratio that has occurred from the period during which humans evolved and their genetic patterns were established (~1:1 to 2:1) until today (~5:1, in worse cases up to 20:1) has increased the occurrence of CVD (Simopoulos 2006). Epidemiologic studies have shown correlations between the n-6/n-3 ratio and CVD supporting this view (Simopoulos 2006). Most likely, both the absolute intake and the ratio are of importance. In the US, it was estimated that 3.5 g LC n-3 PUFA/d was needed for a 2000 kcal diet in order to achieve a decent n-6/n-3 ratio. A better ratio can also be reached by reducing the n-6 fat intake (Hibbeln *et al.* 2006).

[28] Pre-eclampsia is a medical condition where hypertension arises in pregnancy in association with significant amounts of protein in the urine.

Table 2. *Overview of the reviews on LC n-3 PUFA and health cited in the present review.*

End point	Reviews (only Human studies)	Reviews (human and animal studies)
CVD	Balk *et al.* 2006a; Breslow 2006; Hamer and Steptoe 2006; Wang *et al.* 2006	Kristensen *et al.* 2001; London *et al.* 2007
The metabolic syndrome	Mori 2006	Carpentier *et al.* 2006
Cancer	MacLean *et al.* 2006	
Inflammatory diseases		Calder 2006
Brain		
Dementia and macular degeneration	Johnson and Schaefer 2006	
Schizophrenia	Lakhan and Vieira 2008	Mahadik *et al.* 2006
Depressed mood	Appleton *et al.* 2006; Clayton *et al.* 2007	
ADHD	Clayton *et al.* 2007	
Effects during pregnancy	Jensen 2006; Szajewska *et al.* 2006; Wang *et al.* 2006	Cheatham *et al.* 2006

3.2 Marine phospholipids

3.2.1 What are marine phospholipids (PL)?

Phospholipids (PL) are important constituents of biomembranes. Modern biotechnology and engineering make it possible to design and purify marine PL for use in nutrition and drug delivery. Today there are commercial products available for use in the feeding of fish larvae with PL and also for use in cosmetics. The discovery of nutritional and pharmaceutical values of marine PL is spurring on a product development leading to PL of specific structures of high purity (Guo *et al.* 2005).

3.2.2 Health effects

PL are a diverse group of constituents which are involved in many intrinsic applications within the cells. A recent congress on PL and health reported that there is still no clear understanding as to why and how much PL make a difference in human nutrition (Szuhaj and Nieuwenhuyzen, 2003). However, some documented health effects were specially emphasised. Firstly, lecithin (mainly consisting of phosphatidylcholine, PC) plays a role in cholesterol management, in the activity of the lecithin cholesterol acyl transferase (LCAT) enzyme and in the formation of atherosclerotic plaque (Miller 2003). The effects of lecithin on HDL cholesterol were reported to be independent from those of LC n-3 PUFA. Secondly, PC and phosphatidylethanolamine (PE) were postulated to correct PL abnormalities in the brain and to stabilise neuronal membranes. They could hereby be therapeutic agents in Alzheimer's disease and different depressed conditions (Pepeu 2003). Thirdly, dietary sphingolipids[29] were reported to suppress colon carcinogenesis (Schmelz 2003; Merrill *et al.* 2003). The mechanisms with which the lecithin affects cholesterol, LCAT or the atherosclerotic plaque are not completely understood. The primary mechanism of the cholesterol lowering effect has been reported to be due to a decrease in absorption of the dietary cholesterol from the intestine to the bloodstream (Beil and Grundy 1980 and Rampone and Machida 1981 referred to in Miller 2003). It has also been reported that lecithin intake lowers cholesterol by increasing the amount of cholesterol used in the production of bile salt (Mastellone *et al.* 2000; referred to in Miller 2003). The increasing activity of LCAT after a lecithin containing diet is not well understood, but it has been suggested that LCAT may be stimulated by the excess of PL in serum (Miller 2003).

3.2.3 Phospholipids as carriers of nutrients

Marine PL have a higher concentration of LC-n-3 PUFA than the marine neutral lipids. Some studies have shown that the bioavailability of LC n-3 PUFA in PL is better than for TAG´s (Simpolous *et al.* 1992; Boehm *et al.* 1996; Carnielli *et al.* 1998; Lemaitre-Delaunay *et al.* 1999), however there are contradictions in the literature (Sala-Vila *et al.* 2004; Aid *et al.* 2005). Some animal studies have shown a higher absorption of LC n-3 PUFA in specific organs in the animal when feeding with PL compared with TAG (Amate *et al.* 2001ab, Wijendran *et al.* 2002; Werner *et al.* 2004). A recent paper demonstrated that polar lipids from fish (sea bream) attributed more to blood platelet aggregatory properties in a rabbit model system than the neutral lipids (Nasopoulou *et al.* 2007). A review written by Fave *et al.* (2004) has recently reported that it is still questionable whether the species and quantity of PL also affect the bioavailability of fatty acids through digestion and absorption. Lipid droplet size and composition, hydrophobicity of the PL and lipase activity in the digestive tract were factors with a possible impact on the fatty acid bioavailability. Dietary PL supplies were reported to contribute to improved fatty acid bioavailability in subjects with impaired digestive functions (Fave *et al.* 2004). A study by Song *et al.* (1997) demonstrated that DHA in PL were better retained due to less oxidative stress compared to DHA in TAG and ethyl esters. Dietary PL may also be carriers of other nutrients and PL may be used in drug delivery.

[29] Sphingolipids are a class of lipids derived from the aliphatic amino alcohol sphingosine.

Table 3. *Overview of reviews and animal studies on PL and health cited in the present review.*

End point	Reviews	Animal studies
CVD	Miller 2003	Rapone and Machida 1981; Mastelle *et al.* 2000; Werner *et al.* 2004; Nasopoulou *et al.* 2007
Therapeutic effect of brain	Pepeu 2003	Wijendram *et al.* 2002
Suppression of cancer		Merrill *et al.* 2003; Schmelz 2003

3.3 Vitamin D

Fatty fish is one of very few dietary sources of vitamin D. The need for dietary vitamin D varies with the season since vitamin D is synthesised in the skin from 7-dehydrocholesterol by exposure to sunlight. Sun deprivation is common during wintertime in Nordic countries and among the elderly population (Zittermann 2006). The capacity to synthesise vitamin D decreases with age and darker skin, and dietary vitamin D is extra important for these groups. Deficiency and insufficiency of vitamin D have been reported recently to be common in epidemiological studies from Nordic countries such as Denmark, the Netherlands, Sweden and Norway. There is also an ongoing debate on the definition of deficiency (today commonly set at c-25-hydroxyvitamin D <20 nmol/l) (Holick 2007). Circulating levels of 25-hydroxyvitamin D over 75nmol/l or 30ng/mL have been suggested to be required to maximise the health beneficial effects of vitamin D. To achieve this, in the absence of sun exposure, a daily intake of 800-1000 international units (25 µg/day) of vitamin D may be needed (Holick and Chen 2008).

3.3.1 Bone health

Vitamin D deficiency causes rickets in infants and children and osteomalacia in adults. These diseases are associated with decreased bone mineralisation and bone weakness, caused by malabsorption of dietary calcium. In these conditions the plasma concentration of calcidiol, 25OHD, the precursor of calcitriol, is too low for sufficient production of the hormone, which is the key factor for calcium absorption. The malabsorption of calcium causes hypocalcaemia, which stimulates the secretion of parathyroid hormone. Therefore, vitamin D deficiency is associated with secondary hyperparathyroidism (Holick and Chen, 2008).

3.3.2 CVD and metabolic syndrome related diseases

Vitamin D insufficiency has also been related to CVD and type 2 diabetes mellitus (Zittermann 2006). Observational studies in populations with moderate to high risk for CHD have found an inverse relationship between vitamin D and the extent of vascular calcification (Pittas *et al.* 2007; Watson *et al.* 1997) but the association with CHD, stroke and congestive heart failure must be confirmed in further studies (Michos and Melamed 2008). Inverse association between vitamin D status and several risk factors for CVD such as BMI, blood pressure, blood glucose and TAG have been found (Martins 2007).

Experimental data demonstrate that physiological vitamin D actions inhibit pro-inflammatory cytokine release, adhesion molecule release, and proliferation and migration of vascular smooth muscle cells. These are processes that are important for intimal and medial artery calcification. These studies also report that both excess vitamin D, which is rarely seen in the general population, and low levels of calcitriol are associated with vascular and soft tissue calcifications (Zittermann *et al.* 2007). The association of serum levels of vitamin D metabolites with subclinical vascular disease is still controversial. It has been hypothesised that vitamin D deficiency is the potential factor between osteoporosis and vascular calcification, which are associated.

A relatively consistent association between low vitamin D/calcium status and prevalent type 2 diabetes mellitus and the metabolic syndrome has also been reported from observational studies (Michos and Melamed 2008). The evidence from intervention studies with supplementation with vitamin D and/or calcium supplementation is very weak, mainly because of the lack of long-term studies. The combination of vitamin D and calcium supplementation might prevent type II diabetes in subjects with glucose intolerance (Pittas *et al.* 2007).

3.3.3 Inflammatory diseases

25(OH)D3 plasma levels have been found to be inversely related to rheumatoid arthritis[30] showing a circannual rhythm (Cutolo *et al.* 2007). A greater intake of vitamin D has also been associated with lower risk for rheumatoid arthritis and clinical improvement in rheumatoid arthritis patients. This indicates that vitamin D could have an immunosuppressive role although more data is needed. The mechanism thought to be behind this potential effect is that the vitamin D receptor is found in cells of the immune system and could influence the regulation of cells involved in autoimmune disease (Cutolo *et al.* 2007).

[30] Rheumatoid arthritis is a chronic, systemic autoimmune disorder that causes the immune system to attack the joints, where it causes inflammation (arthritis) and destruction.

3.3.4. Cancer

Experimental evidence suggests that vitamin D may reduce the risk for cancer through regulation of cellular proliferation and differentiation and inhibition of angiogenesis[31]. The anticancer effects are mainly attributed to the hormonal form of vitamin D (calcitriol). Epidemiological studies have shown that vitamin D prevents several forms of cancer such us colorectal, colon, breast, ovarian, endometrial, prostate and lymphoma cancer (Garland 2006). The strongest correlation has been shown for colorectal cancer (Gorham 2005). Mortality in prostate cancer is, for example, inversely related to UV radiation. Cancer at other sites have also been related to higher death rates in subjects with inadequate vitamin D although the beneficial effects of vitamin D on cancer need to be further evaluated in prospective studies (Ali and Vaidya 2007).

3.3.5 Brain

Recent research has indicated that vitamin D could influence brain function. Calcitriol seems to be important for normal brain development and to have neuroprotective functions (McGrath 2004, Kalueff 2004). A small study found that vitamin D was more effective than phototherapy in the treatment of seasonal affective disorder (Gloth 1999).

3.3.6 Pregnancy

Vitamin D deficiency is proposed to be a risk factor for maternal pre-eclampsia. The explanation could be the beneficial effects of vitamin D on the elastic wall of blood vessels (Michos and Melamed 2008). Adequate vitamin D concentrations are also necessary during pregnancy to ensure appropriate maternal responses to the calcium demands of the foetus and neonatal handling of calcium (Specker 2004). The data on weather vitamin D supplementation in women at high risk for vitamin D deficiency could improve maternal weight gain and foetal growth is inconclusive (Specker 2004).

---

[31] Angiogenesis is a physiological process involving the growth of new blood vessels from pre-existing vessels.

Table 4. *Overview of the reviews on vitamin D and human health cited in the present review.*

End point	Reviews (human studies only)
Bone health	Cranney *et al.* 2007
Inflammatory diseases	Cutolo *et al.* 2007
CVD/metabolic syndrome related	Watson *et al.* 1997; Zittermann 2006; Pittas *et al.* 2007; Zittermann *et al.* 2007; Michos and Melamed 2008; Martins *et al.* 2007
Cancer	Gorham *et al.* 2005; Garland *et al.* 2006; Ali and Vaidya 2007
Brain	McGrath *et al.* 2004; Kalueff *et al.* 2004; Gloth *et al.* 1999
Pregnancy	Michos and Melamed 2008

4. Consumption of seafood-derived proteins, peptides, free amino acids and trace elements

4.1 Introduction

The nutritional value or quality of structurally different proteins varies and is governed by amino acid composition, ratios of essential amino acids, susceptibility to hydrolysis during digestion, source, and the effects of processing. The nutritional value of proteins from different food sources, including seafood, are extensively reviewed by Friedman (Friedman 1996). Fish muscle protein is generally rich in lysine, the sulphur-containing amino acids and threonine, which are the limiting amino acid in children's cereal-based diets in developing countries. Therefore, increasing the proportion of marine fish in the diet of people where cereals are the main protein source is an effective way to enhance the nutritional value of food and improve the nutritional status of the developing countries.

Recently there has been an increasing focus on the more specific role of seafood proteins in human health; also within the western population. This applies both to intact proteins and protein hydrolysates, the latter being commonly prepared chemically and enzymatically, e.g. from small fish species or from fish by-products. As such, the health beneficial effects of protein hydrolysates could be discussed also in the second section of this chapter 'Health beneficial effects achieved from seafood by-product utilisation and food fortification'. However, since dietary proteins are enzymatically and chemically hydrolysed also *in vivo* e.g. during gastrointestinal (GI) digestion, tentative health effects from seafood peptides are also pertinent to this section. In fact, a difficulty that would need attention lies in understanding whether the effects of ingested peptides (i.e. hydrolysates) differ from those that arise from peptides formed in the GI tract after ingesting intact proteins. Thus, it is hard to know whether the peptides or proteins should be ascribed an observed effect. Another aspect of seafood peptides and seafood proteins that deserves attention in this chapter is the fact that they have rarely been tested in animal or human models in completely lipid-free forms. Thus, it cannot be fully excluded that some of the effects reported regarding seafood proteins/peptides originate from the LC n-3 PUFA. It is certainly easier to interpret the effects of those amino acids in seafood that are free from the start, i.e. taurine, glutamine, proline, glycine, alanine and arginine. Here, one has no interference from the potential benefits of peptides/amino acids formed from ex vivo hydrolysis or GI metabolism of proteins.

4.2 Fish proteins

4.2.1 Uptake of dietary proteins

Although attention has focused on the benefits of fish oils in heart disease, a significant proportion of dietary fish intake is comprised of protein. This is especially true of lean, white fish such as cod. Measurements of proximate composition of cod give values of 15-17% protein and 1-2.5% fat depending on the season (Holdway and Beamish 1984).

Proteins are partially or fully denatured by normal cooking temperatures, but hydrolysis to small peptides and component amino acids is largely undertaken by stomach acid and digestive enzymes. Animal studies (Curtis and Gill 1992) have shown that both whole proteins and large polypeptides survive degradation in the stomach. Similarly, systemic exposure to proteins and peptides derived from food are the basis for antibody mediated food allergies (Chambers *et al.* 2004). Therefore, there is potential for the uptake of bioactive proteins and peptides directly from the diet.

4.2.2 Effects on cardiac risk factors - high blood pressure and TAG levels

The positive health effects of a diet containing purified fish protein have been documented in animal models of CVD. Ait-Yahia *et al.* (2003) and Boukortt (2004) showed that a diet containing commercially prepared, virtually fat-free fish protein lowered blood pressure in hypertensive rats. High blood pressure is a major risk factor for CVD.

In a study by Demonty *et al.* (2003), the combined effect of de-fatted cod protein/casein and fish oil/beef tallow on TAG-metabolism modulation was studied in rats. Hepatic TAG concentrations and TAG-secretion rates were reduced in rats fed cod protein compared to casein. The protein source did not exert an independent effect on triglyceridaemia, but the combination of cod protein+menhaden oil resulted in 50% lower plasma TAG levels than casein beef tallow. Casein+menhaden oil did not significantly reduce triglyceridaemia compared to casein beef tallow. Menhaden oil and cod protein also lowered plasma cholesterol concentrations in comparison with beef tallow and casein, respectively. The latter was linked to reduced hepatic cholesterol levels in rats fed cod proteins compared to those fed casein. Thus, this single study indicated independent beneficial effects from both cod protein and menhaden oil on lipid metabolism in rats.

4.2.3 Effects on obesity, metabolic syndrome and type II diabetes

A major and growing group of health problems are those associated with obesity. In addition to obesity being a major risk factor for atherosclerosis and high blood pressure, both contributors to the development of CVD, obesity is strongly associated with metabolic syndrome and type II diabetes. The metabolic syndrome is a complex condition comprised of changes in circulating lipids, reduced glucose tolerance and increased insulin resistance. Unchecked metabolic syndrome can develop into type II diabetes characterised by high levels of circulating lipids, high blood glucose and complete insensitivity to insulin. Arterial disease, cardiac hypertrophy[32] and reduced heart function are well-characterised complications of type II diabetes. Animal studies and one human study have shown that fish protein as a major component of the diet might have the potential to improve health in metabolic syndrome and contribute to the prevention of type II diabetes (Lavigne *et al.* 2000, 2001; Papakonstantinou *et al.* 2005; Ouellet *et al.* 2007).

Ouellet *et al.* (2007) compared the effect of cod proteins with other animal proteins (beef, pork, veal, egg, milk) regarding their effects on insulin sensitivity in insulin-resistant human subjects. Diets were formulated to only differ in the protein source while fatty acids and fibres were identical. Cod proteins significantly improved insulin sensitivity and had a strong tendency to cause better β-cell function. Cod proteins could thereby contribute to prevention of type II diabetes by reducing the metabolic compliance related to insulin resistance.

In animal model studies in which the dietary protein source was freeze-dried, defatted cod fillets, the animals showed improvements in glucose and lipid metabolism (Hurley *et al.* 1995), improved glucose tolerance and insulin sensitivity (Lavigne *et al.* 2000; Tremblay *et al.* 2005) and improved insulin signalling leading to glucose uptake (Trembley *et al.* 2003, 2007). When studying rats fed on a high-fat diet, it was found that the use of casein and soy protein as protein sources caused severe whole body and skeletal muscle insulin resistance to develop (Lavigne *et al.* 2001). However, feeding defatted cod protein fully prevented this development of insulin resistance. No reductions in body weight gain, adipose tissue accretion, or expression of TNF-α in fat and muscle were seen in these rats. This study also showed

[32] Hypertrophy is the increase of the size of an organ or a select area of the tissue.

that L6 myocytes exposed to cod protein-derived amino acids showed greater rates of insulin-stimulated glucose uptake compared with cells incubated with casein- or soy protein-derived amino acids. These data demonstrate that cod protein intake prevents obesity-induced muscle insulin resistance in high fat-fed obese rats at least partly via a direct action of amino acids on insulin-stimulated glucose uptake in skeletal muscle cells.

A review (Halton and Hu 2004) of 15 randomised, controlled trials into the effects of high-protein diets on body weight and cardiovascular risk factors showed a close association between both total protein intake and type of protein consumed and weight control. High-protein diets, especially non-meat derived proteins, promote weight loss, improve plasma lipid status and have beneficial effects on insulin sensitivity.

Taken together, these results indicate that a diet rich in lean fish protein has the potential to reduce cardiac risk factors and improve metabolic function both indirectly by aiding in the control of obesity and directly via one or several active components of fish protein.

Table 5. *Overview of reviews, human intervention studies and animal studies on fish proteins and human health cited in the present review.*

End point	Reviews	Interventions	Animal studies
High blood pressure			Bourkott *et al.* 2004; Ait-Yahia *et al.* 2003
Lipid metabolism			Demonty *et al.* 2003
Obesity/metabolic syndrome	Halton and Hu 2004		
Glucose and lipid metabolism			Bourkott *et al.* 2004; Ait-Yahia *et al.* 2003; Hurley *et al.* 1995; Lavigne *et al.* 2000; Tremblay *et al.* 2003
Insulin sensitivity	Tremblay *et al.* 2007	Ouellet *et al.* 2007	Lavigne *et al.* 2001; Papakonstantinou *et al.* 2005;Tremblay *et al.* 2005

4.3 Fish peptides and hydrolysates

4.3.1 Introduction

Fish protein hydrolysates obtained by enzymatic hydrolysis of various fish tissues are rich sources of bioactive peptides. However, they have been less investigated than for instance milk peptides, which have been subjected to considerable research for several decades (Meiser and Schlimme 1996). As stated above, bioactive peptides may principally occur in two very different manners. In living organisms bioactive peptides are released as a result of activation of regulatory enzymes which split off specific peptides from specific proteins. In this case the peptides have predestined biological functions. However, identical peptides may also occur *ex vivo*, more or less randomly, as a result of autolytic digestion of a mixture of proteins by various proteolytic enzymes. Obviously the probability of obtaining small bioactive peptides (di- or three-peptides) as a result of random digestion is significant. Recently, a review on bioactive compounds from marine processing by-products was presented by Kim and Mendis (2006).

4.3.2 Effects as ACE inhibitors - reduction of high blood pressure

Matsumura *et al.* (1993) found small peptides in an autolysate of bonito bowels which inhibited ACE-converting enzymes *in vitro*. ACE inhibiting peptides have also been isolated from fish scales, hydrolysates of cod head and shrimp (both *Pandalus borealis* and *Acetes chinensis*) as well as from fish sauce (Fahmi *et al.* 2004; Bordenave *et al.* 2002; He *et al.* 2006; Ichimura *et al.* 2003). Recently Jung *et al.* (2006) isolated an efficient ACE-inhibitory peptide from a protein hydrolysate of yellowfin sole *(Limanda aspera)*. The peptide, containing 11 amino acid residues, had an *in vivo* antihypertensive effect after oral administration of spontaneously hypertensive rats.

One randomised double-blind placebo-controlled study has been conducted for 13 weeks on 63 human subjects to determine the antihypertensive effect of a vegetable drink (195 g/day) fortified with sardine protein hydrolysates (0.5 g/day) containing a dipeptide, Valyl-Tyrosine (0.4 mg/day) (Kawasaki *et al.* 2002). Controls got the same drink without peptides. The subjects had mild hypertension, high to normal blood pressure or normal blood pressure. In the two former groups, both systolic and diastolic blood pressure was significantly reduced. No changes were seen in the control group or group with normal blood pressure. An excessive ingestion test was performed on 25 subjects with hypertension, mild hypertension, high to normal blood pressure, and normal blood pressure by giving 585 g (3 times the recommended amount of intake) of the test drink for 14 days in a row. A significant decrease of blood pressure was observed in the hypertension, mild hypertension and high to normal blood pressure groups. In the groups with normal blood pressure, the excessive ingestion of the test drink did not affect blood pressure. In these two studies, physical check-ups and biochemical analyses of blood and urine were also conducted in all subjects, and no abnormalities were observed. These results suggest that the test drink containing sardine protein hydrolysates exhibited the antihypertensive effect only in the subjects with mild hypertension or high to normal blood pressure.

4.3.3 Immuno-stimulants

In vitro experiments showed that small peptides from fish sauce (0.5-3 kDa) stimulated proliferation of human monocytes *in vitro* (Thonthtai and Gildberg 2005), whereas small acid peptides in Atlantic cod stomach autolysates enhanced superoxide anion production in Atlantic salmon leucocytes (Gildberg *et al.* 1996). Phagocytic activity in mice macrophages was enhanced after feeding with Pacific whiting *(Merluccius productus)* protein hydrolysates (Duarte *et al.* 2006).

4.3.4 Antioxidative activity

Antioxidative peptides have been found in hydrolysates of Alaska pollack skin (Kim *et al.* 2001). Specific peptides with a Gly-Pro-Hyp C-terminus revealed *in vitro* antioxidative activity determined by the thiobarbituric acid reactive substances (TBARS) method and improved cell survival after addition to cultures of rat liver cells. Sathivel *et al.* (2003) found that protein hydrolysates of Atlantic herring by-products revealed an *in vitro* antioxidative capacity by inhibiting the peroxide formation in incubation mixtures containing linoleic acid.

4.3.5 Anti-carcinogenic effects

Recently Picot *et al.* (2006) *in vitro* achieved proliferation inhibition in two human breast cancer cell lines by adding various fish protein hydrolysates made from cod, plaice and salmon to the cell cultures.

4.3.6 Anti-anaemia activity

A protein hydrolysate made from *Saurida elongate* revealed *in vivo* anti-anaemia activity in mice where blood loss had been experimentally induced. Mice fed fish protein hydrolysate recovered more rapidly from the anaemia than control groups (Dong *et al.* 2005).

Table 6. *Overview of reviews, human interventions and cell/animal studies related to bioactive peptides from fish protein hydrolysates cited in present review.*

End point	Reviews	Interventions	Cell/animal studies
High blood pressure	Meisel and Schlimme 1996; Thonthai and Gildberg 2005; Kim and Mendis 2006	Kawasaki *et al.* 2002	Matsumura *et al.* 1993; Bordenave *et al.* 2002; Ichimura *et al.* 2003; Fahmi *et al.* 2004; He *et al.* 2006; Jung 2006
Low immune response	Thonthai and Gildberg 2005; Kim and Mendis 2006; Meisel and Schlimme 1996		Gildberg *et al.* 1996; Duarte *et al.* 2006
Cancer	Kim and Mendis 2006; Thonthai and Gildberg 2005		Kim *et al.* 2001; Sathivel 2003; Picot *et al.* 2006
Anaemia			Dong *et al.* 2005

4.4 Selected amino acids in fish

4.4.1 Introduction

It is generally accepted that the relative concentration of essential amino acids is the major factor determining the nutritional value of food proteins. Seafood muscles are rich in water-soluble components; among which are free amino acids. Free amino acids provide the main components of non-protein nitrogen compounds and are known to influence the taste of food significantly. Major free amino acids of seafood muscle are taurine, glutamine, proline, glycine, alanine and arginine. When by-products are to be used as a source of seasoning, the spectrum and concentration of free amino acids would greatly influence their taste and quality. However, preparation of seafood causes losses of water-soluble compounds and the health effects emanating from such components (free amino acids, minerals and trace elements) may thus be expected to be elevated when consuming less prepared foods.

Although data are still limited, it is also likely that the interactions of LC n-3 PUFAs and other nutrients, including amino acids found in fish, play important roles in physiological and biological metabolisms related to cardiovascular function. In this section, two amino acids that are suggested to have effects on human health beyond the provision of basic nutrients when consumed in amounts obtainable from the diet, are highlighted; arginine and taurine.

4.4.2 Arginine

Fish (and meat) proteins contain relatively high amounts of arginine (1-1.2 g/100 g muscle) (FAO 1970). In shellfish, even higher levels are found (1.4-1.8 g/100 g). Arginine, which is found both in the free (mainly shellfish), and bound amino acid pool, is an essential precursor for the synthesis of proteins and other molecules with biological importance. Arginine administration has been suggested to be important in many pathophysiological[33] conditions in humans (Niittynen *et al.* 1999). When consumption of fish protein (grilled cod) was compared to meat (beef) in a limited number of healthy humans (6 males), an enhanced increase in plasma arginine was demonstrated (Soucy and LeBlanc 1998). A reduced secretion of postprandial insulin was also recorded in this experiment.

The possible effects of activating the l-arginine-nitric oxide pathway have been reviewed by Brown and Hu (2001). Synergistic effects of LC n-3 PUFAs on eicosanoid balance and the l-arginine-nitric oxide pathway have been suggested to provide additional cardioprotective effects (He and Daviglus 2005).

4.4.3 Taurine

Differences in muscle osmolality, e.g. between marine and non-marine animals, are mainly due to nitrogenous solutes such as certain free amino acids, among them taurine (Abe 2000). Seafood, especially invertebrates such as molluscs and crustaceans are high in taurine (300-800 mg per 100 g edible portion) (Roe and Weston 1965; Zhao *et al.* 1998; Spitze *et al.* 2003; Dragnes *et al.* 2008) and by-products of seafood may thus serve as raw materials for taurine-containing foods (Elvevoll 2005). The taurine concentrations in a variety of common seafood and their corresponding raw materials have recently been reported (Dragnes *et al.* 2008). The amount of taurine in fish is variable. Taurine concentrations (per 100g raw fillets) of farmed Atlantic salmon (*Salmo salar*) (94 ± 16 mg), cod (*Gadus morhua L.*) (120 ± 21 mg), saithe (*Pollachius virens*) (162 ± 25 mg), haddock (*Melanogrammus aeglefinus*) (57 ± 6 mg) and products of mackerel (*Scomber scombrus*), Alaska pollack (*Pollachius pollachius*), yellowfin tuna (*Thynnus albacares*), shrimps (*Pandalus borealis*) or blue mussel (*Mytilus edulis*) are reported.

Humans have a limited ability to biosynthesise taurine and it may be regarded as conditionally essential as its physiological concentration can be partly regulated endogenously (Stapleton *et al.* 1997). Taurine is synthesised from cysteine via the

[33] Pathophysiological is the disturbance of normal mechanical, physical, and biochemical functions.

sequential actions of cysteine dioxygenase (CDO), which gives rise to cysteine sulphinate, and cysteine sulphinate decarboxylase (CSD), which decarboxylates cysteine sulphinate to hypotaurine. Hypotaurine is further oxidised to taurine. The capacity for taurine biosynthesis varies between species. As an example, compared with the *in vitro* CSD activity found in rat, the activity in man, primates and the cat are very low, which is believed to be the rate-limiting enzyme responsible for the formation of taurine from cysteine (Bouckenooghe *et al.* 2006).

Taurine is thus an end product of sulphur amino acid metabolism and considerations of the need for supplementation or dietary changes have to include consumption data on methionine and cysteine. Methionine is the only essential sulphur amino acid (SAA) and can provide sulphur for cysteine and taurine synthesis. Animal protein is generally considered to be a better source of SAA than vegetable protein. There is apparent consensus concerning normal SAA requirements. The classical experiments of Rose (1976) have been used and reproduced by WHO (FAO/WHO/UNU 1985) and several other authors (Van de Poll 2006) and an SAA intake of 13 mg/kg per 24 h for healthy adults is still recommended.

Fish muscle consumption is reported to result in increased concentrations of serum taurine when compared with beef and chicken muscle (Uhe *et al.* 1992). Humans on a diet high in seafood are reported to be high in serum taurine (Zhao *et al.* 1998; Kibayashi 2000; Kim *et al.* 2003) whereas humans on diets relatively low in seafood are reported to be low in serum taurine (Brøns *et al.* 2004). Human urinary excretion of taurine is also known as a marker for seafood consumption (Biosca *et al.* 1990; Kim *et al.* 2003).

4.4.3.1 Effects of taurine on CVD and CVD risk markers

The beneficial effects of dietary supplementation with taurine, in animal and human models, have been reviewed (Militante and Lombardini 2004). A reduced CVD risk through taurine, alone or in combination with LC n-3 PUFA in seafood has been put forward in several papers (Mizushima *et al.* 1997; Yamori *et al.* 1994, 2001, 2006) which have been reviewed by Yamori *et al.* (2004).

The amino acid taurine has been shown to have several positive effects on the cardiovascular system and a broad review is presented by Niittynen *et al.* (1999). Taurine was shown to have antioxidant activity and reduced the production of inflammatory mediators in a mouse model (Oudit *et al.* 2004). Taurine has also been shown to lower blood pressure in borderline hypertensive patients. It has further been reported that taurine can improve cardiac performance, reduce blood cholesterol values and suppress platelet aggregation. Taurine-supplemented diets have been shown to improve bile faecal excretion and to suppress the development of atherosclerotic lesions in mice and rabbits (Balkan *et al.* 2004).

One of the well-documented biological activities of taurine is bile salt formation. Taurine (and glycine) conjugates with cholesterol derivates to form taurocholate (and glycocholate). Taurocholate is the major bile salt that extracts cholesterol from plasma in humans and a decreased taurine content is associated with a lower cholesterol extraction and subsequent accumulation and increased risk of atherosclerosis. The anti-atherosclerotic effects of taurine have been studied in different hypercholesterol-aemic and hyperlipidaemic animals, but the exact mechanism of action is still unclear. Reviewed human trials (supplemented with high amounts 3-6 g per day) and animal intervention trials have also demonstrated that taurine alone has beneficial effects on serum lipids in rats, mice, rabbits and humans (Militante and Lambordini 2004). In contrast, a recent human intervention trial with supplementation of 1.5 grams taurine/day (Brøns *et al.* 2004) showed no effects on blood lipids in overweight men with a genetic predisposition for type II diabetes mellitus.

Recently, a human intervention study on the combined effects of LC n-3 PUFA and taurine has been performed. Healthy volunteers, a total of 80 individuals, were recruited and divided in two groups to attend a seven-week double-blind and parallel intervention trial. One group received fish pâté enriched with n-3 (1.1 g per day) and the second an identical pâté enriched with both LC n-3 PUFA and taurine (425 mg/day). Total cholesterol, LDL-cholesterol and Apo B decreased significantly more in the LC n-3 PUFA+taurine compared to the LC n-3 PUFA group. Also a significant within group enhancement of HDL-cholesterol was demonstrated in the LC n-3 PUFA+taurine group (Elvevoll *et al.* 2008).

4.4.3.2 Taurine and mental health

Taurine is regarded as important in membrane stabilisation and in the development of the central nervous system and the retina (Bouckenooghe *et al.* 2006). An interesting correlation between seafood consumption and social development among young adults and children has been published and has recently been widely discussed. Recently, a paper on foetal neurodevelopment as being affected by taurine has also been published (Hibbeln *et al.* 2007). These kinds of findings have to date been attributed to LC n-3 PUFA alone and pinpoints the fact that further research on human subjects is needed to clarify the possible contribution of marine 'non omega-3 compounds'.

4.4.3.3 Taurine and sports

Taurine-containing energy or sports drinks have been on the European market for about a decade, and research on the individual constituents of these drinks indicates an improvement in cognitive performance resulting from consumption of such drinks. The daily intake of taurine in Western Europe, and the safety of drinks enriched with taurine, were reviewed by The European Scientific Committee on Food (EFSA) in 1999 (EFSA Scientific Committee on Food 1999). EFSA concluded that toxicological studies did not reveal any indication for a genotoxic, carcinogenic or teratogenic potential of taurine. In addition, the committee requested additional adequate systematic studies on subchronic as well as chronic toxicity/carcinogenicity of taurine as the available data are insufficient to establish an upper safe level for daily intake. This task was also revisited by the same committee in 2003, when a new 13-week study on rats provided data on possible behavioural effects and impaired motor performance resulting from taurine supplementation. This time the committee requested focused neurological studies and made an exception to the high taurine content of 'energy' drinks allowing an estimated acute intake of taurine of up to 3 g/day from consumption. This can be compared with the highest estimated intake of taurine from naturally occurring sources in the European diet which is 400 mg/day. The risks and benefits of enriching the diet with taurine, L glutamine and L arginine through supplements or functional food and beverages is also the theme in a recent review by Shao & Hathcock (2008). Here the absence of a systematic pattern of adverse effects in humans in response to orally administered taurine, and thus, the insufficient data to establish an upper safe level (the Observed Safe Level, OSL) is highlighted. Based on data from the available published human clinical trials, the authors conclude that the evidence for the absence of adverse effects is strong for taurine at supplemental intakes up to 3 g/d which level is identified as the OSL for normal healthy adults.

Table 7. *Overview of reviews, observational studies, human interventions and animal studies dealing with amino acids cited in the present review.*

End point	Reviews	Observational studies	Interventions	Animal studies
CVD	Niittynen *et al.* 1999; Brown and Hu 2001; Bouckenooghe *et al.* 2006			Oudit *et al.* 2004
CVD/IHD mortality		Yamori *et al.* 1994, 2001, 2006	Yamori *et al.* 2001, 2006	
CVD	Niittynen *et al.*; Yamori *et al.* 2004	Mizushima *et al.* 1997	Brøns *et al.* 2004; Elvevoll *et al.* 2008	
Atherosclerosis	Niittynen *et al.* 1999; Brown and Hu 2001; Bouckenooghe *et al.* 2006			Militante and Lombardini 2004; Balkan *et al.* 2004
Blood lipids	Bouckenooghe *et al.* 2006		Brøns *et al.* 2004; Elvevoll *et al.* 2008	Militante and Lombardini 2004
Inflammation	Niittynen *et al.* 1999; Bouckenooghe *et al.* 2006			
Oxidative stress	Brown and Hu 2001; Bouckenooghe *et al.* 2006			Oudit *et al.* 2004; Balkan *et al.* 2004
Diabetes II	Bouckenooghe *et al.* 2006		Brøns *et al.* 2004	
Inflammatory diseases	Niittynen *et al.* 1999; Bouckenooghe *et al.* 2006			
Brain	Bouckenooghe *et al.* 2006			
Cognitive development		Hibbeln *et al.* 2007		
Other diseases	Roe and Weston 1965; Bouckenooghe *et al.* 2006			

4.5 Selenium (Se)

4.5.1. Introduction

Several review papers have appeared (Rayman 2002, 2005; Whanger 2004; Combs 2005) addressing the health beneficial effects of Se in relation to CVD, oxidative stress conditions, the immune system, viral infections, reproduction, thyroid function, mood and cancer. Positive effects on immune system stimulation and reductions of cancer incidence/cancer mortality have been obtained in some cases where the Se intake has been above those levels supposed to be required for repletion[34] of the seleno enzymes.

Excluding cereals, seafood is one of the food commodities relatively rich in Se (0.2-0.5 mg Se/kg fish muscle tissue). The Se content in farmed fish is at the lower end (0.2 mg/kg fish muscle tissue) but data are scarce. The contribution of fish to the total food intake of Se is approx. 15-20%. Se is an essential element and it forms a part of at least eleven seleno-proteins in two groups of seleno-enzymes in the human body: glutathione peroxidases (GPx) and iodothyronine deiodinases. Se is beneficial at lower concentrations, whereas at higher concentrations it becomes toxic. The range between deficiency, essentiality and toxicity, however, is rather narrow.

The Scientific Committee on Food (SCF) from the European Commission (2000) recommended that Se intake should not exceed 300 μg Se per day. The Dietary Reference Intake[35] established by National Academy of Sciences (2000) in USA is 55 μg Se/day for adult men and women. A so-called Population Reference Intake (PRI) of 55 μg Se/day was established by the SCF (1993). However, in some European countries different PRI values are recommended varying from 30-150 μg Se/day. The Se intake based upon different studies in Europe, reviewed by SCF (2000), is in Belgium 28-61 μg/day, Denmark 41-47 μg/day, Finland 100-110 μg/day, France 29-43 μg/day, United Kingdom 63 μg/day, the Netherlands 40-54 μg/day, Norway 28-89 μg/day, Spain 79μg/day and Sweden 24-35 μg/day. Rayman (2005) argued that the current plasma and serum Se concentrations do not allow maximal expression of plasma glutathione peroxidase (GPx), the criterion adopted by most authorities when setting their dietary reference values for Se.

4.5.2 Bioavailability of Se from fish

The bioavailability and metabolic fate of Se from dietary fish in humans have not been studied extensively. A cross-sectional study (Hagmar *et al.* 1998) among coastal

[34] The condition of being fully supplied or completely filled.
[35] Also called Recommended Daily Allowance or Recommended Dietary Intake.

fishermen and inland men from Latvia (in total 68 men, 24-79 years) was carried out in order to investigate the relationships between fish intake and different markers of Se status and thyroid hormone function. The number of fish meals per month was correlated with plasma Se, selenoprotein P and GPx. The mean plasma Se level in the subjects with high fish intake was 81% higher than in those with the lowest intake.

An intervention study (Fox *et al.* 2004) was carried out to measure the bioavailability of Se from trout, yeast and selenate. The study had a parallel, randomised, reference substance controlled design and was carried out with 35 volunteers in the Netherlands and United Kingdom. Apparent absorption of Se from cooked or salted enzymatic ripened trout (88 ± 5% respectively 90 ± 3%) was similar to selenate (93 ± 4%). It also showed that there was no difference between the two fish processing methods used. Apparent absorption of yeast Se (54 ± 7%) was significantly lower than from Se in trout and selenate. However, Se retention from trout (86%) was significantly higher than selenate (60%). The retention of Se from yeast (59%) was lower.

In another study (Fox *et al.* 2005), wheat, garlic and cod intrinsically labelled with Se-77 or Se-82 stable isotopes were consumed in random order by 14 adults. The minimum wash-out period was six weeks between each test meal. Se absorption was significantly higher from wheat (81 ± 3%) and garlic (78 ± 14%) than from cod (56 ± 4%). The form of Se, and the food constituents with which it appears are suggested to be key determinants of post-absorptive metabolism.

4.5.3 Speciation of selenium in fish

From a nutritional and health beneficial point of view, it is even more important to know the chemical form in which Se is present in the edible part of the fish. The chemical form determines the degree of bioavailability and bioactivity after absorption. Although the methodologies for speciation trace element analysis have improved over the last ten years the speciation of Se in fish is still not optimal. Poor extraction recovery of Se from the raw material is one of the obstacles in that respect.

Early efforts to determine Se speciation in fish (plaice, cod, mackerel, herring) were performed by Åkesson and Skrikumar (1994), Shen *et al.* (1997), Crews *et al.* (1998) and by Önning and Bergdahl (1999) using various forms of size-exclusion chromatography with off- or on-line detection of Se. These studies demonstrated that the extraction of Se from the edible part was poor (less than 30% for some species). The soluble fraction (mainly water-soluble proteins and low molecular weight (LMW) compounds) from plaice and mackerel contained a high proportion of LMW Se compounds while high molecular weight Se compounds were present in cod.

More recently selenomethionine and selenocystine were identified in enzymatic hydrolysates from Antarctic krill (Siwek *et al.* 2005). Several other studies (Quijano *et al.* 2000; Moreno *et al.* 2004; Cabanero *et al.* 2005) identified selenomethionine as the main Se component in the soluble fractions of sardine, swordfish, tuna, trout, krill, oyster and mussel. $TrimethylSe^{+}$ and inorganic Se were also found in some species.

A complete review of Se speciation from food source to metabolites has recently been published by Dumont *et al.* (2006). This review shows that selenomethionine and selenocystine are the main Se components in Brazil nuts. Selenomethionine is the major component in cereals. In garlic and broccoli γ-glutamyl-Se-methylselenocysteine and Se-methylselenocysteine are predominantly present. Se yeast supplements contain mainly selenomethionine.

4.5.4. Effects on cancer

In rat studies, supplementation of an Se garlic diet at different levels (1-3 µg Se/g) consistently caused a lower tissue Se accumulation when compared to Se yeast (Clement *et al.* 2000). On the other hand, Se-garlic was significantly more effective in suppressing the development of premalignant lesions[36] and the formation of adenocarcinomas[37] in the mammary gland of carcinogen-treated rats.

In another study (Finnley *et al.* 2000) diets with various amounts (up to 2.0 µg Se/g diet) and forms of Se (selenate and selenised broccoli) were given to rats. Supranutritional amounts of Se supplied as broccoli significantly decreased the incidence of aberrant crypts and aberrant crypt foci[38] being preneoplastic lesions[39] indicative for colon cancer.

Prospective studies of Se and cancer in humans, reviewed by Knekt (2002), showed that in approximately 50 out of 72 studies, a lower cancer risk was associated with higher Se intake. The strongest evidence for a beneficial effect of Se appears to be related to lung cancer, oesophageal and gastric-cardia cancers and, most notably, prostate cancer.

The strongest evidence of the efficacy of Se as a cancer prevention agent, particularly for prostate cancer, is provided by the Nutritional Prevention of Cancer trial (Clark *et al.* 1996). Subjects with a history of non-melanoma skin cancer were treated for 4.5 years with yeast-Se (200 µg Se/day) and the follow-up time was 6.5 years.

[36] Abnormality in a tissue area which is just a step away from cancer.
[37] A form of carcinoma that originates in glandular tissue.
[38] Putative precursors of colon cancer.
[39] Preceding the development of a tumor abnormal tissue.

Fifty percent lower total cancer mortality was found and 37% lower total cancer incidence.

Further randomised clinical trials using defined Se compounds are needed for confirmation of the above-mentioned clinical results. One such study is currently in progress in the USA, the Se and Vitamin E Cancer Prevention Trial (SELECT). Planned as a 12-year trial involving 400 sites and an enrolment of 32,400 men, SELECT will test the hypothesis that selenomethionine (200 µg/day) and or vitamin E supplementation can reduce the risk of prostate cancer (Klein *et al.* 2000)

In order to test the possible role of Se in cancer prevention, a group of researchers from Denmark, Sweden and the United K.ingdom have designed the Prevention of Cancer by Intervention with Selenium (PRECISE) study. Over a period of 8 years a planned total of 42000 Europeans will be randomised to supplementation with yeast Se at 100, 200 and 300 µg/day or placebo.

A number of mechanisms have been suggested to explain the anti-cancer effects of Se. Although it is fairly well accepted that methylselenol (CH_3SeH) is involved in the anti-cancer effects of Se at supra-nutritional doses, evidence is growing that the seleno-enzymes do play a role, particularly at nutritional levels of intake. CH_3SeH can be formed by methylation of H_2Se as part of the Se excretory pathway or formed from a storage form of Se in plants (broccoli and garlic), i.e. γ-glutamyl-selenomethyl-selenocysteine. Metabolism removes the γ-glutamyl group to give selenomethyl-selenocysteine which is converted to give CH_3SeH (Figure 1).

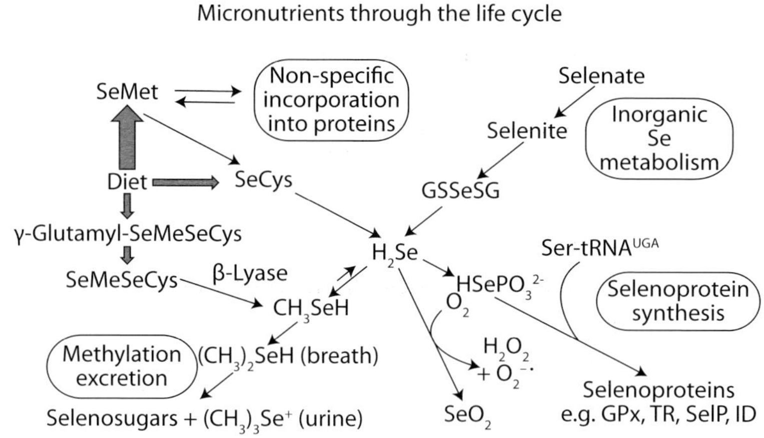

Figure 1. *The metabolism of dietary forms of selenium. SeMet, selenomethionine; SeCys, selenocysteine; SeMeSeCys, selenomethyl-SeCys; GSSeSG, selenodiglutathione; GPx, glutathione peroxidase; TR, thioredoxin reductases; SelP, selenoprotein P; ID, iodothyronine deiodinases (Adapted from Rayman 2005).*

Table 8. *Overview of reviews, observational studies, intervention studies and animal studies on selenium and human health cited in the present review.*

End point	Reviews	Observational studies	Interventions	Animal studies
CVD, immune system, viral infections, reproduction, thyroid function, mood and cancer	Rayman 2002, 2005; Whanger 2004; Combs 2005			
Cancer	Knekt 2000		Clark *et al.* 1996	
Mammary gland rats				Clement 2000
Colon cancer				Finnley 2000

5. Consumption of compounds achieved by seafood by-product utilisation and food fortification

5.1 Introduction

By-products are identified as leftovers that are not ordinarily saleable, but which can be recycled after treatment. By-products from fish mainly consist of viscera (liver, roe, stomachs, etc.), heads, backbones, cuts and rejected fish from processing. Shellfish and crustacean shells are another important source. The by-products are generated when the fish/shellfish is gutted, headed and further processed - either on-board in fishing vessels or in processing plants on shore. According to FAOSTAT (2001), discards from the world's fisheries in 2001 exceeded 20 million tons, equivalent to 25% of the total production of marine capture fisheries. The majority of fisheries by-products are presently employed to produce fish oil, fish meal, fertiliser, pet food and fish silage (Kim and Mendis 2006). However, most of these recycled products possess low economic value. A large part is also still dumped into the sea, mainly from the fishing fleet. Recent studies have identified a number of bioactive compounds, e.g. from shellfish and crustacean shells, fish bone and cartilage (Kim and Mendis 2006). This is highly interesting as it has been estimated that if we succeed in utilising more seafood by-products as food for humans and as ingredients in foodstuff, health foods, nutraceuticals, pharmacy, cosmetics, etc., the added value may increase 5-fold (www.rubin.no).

5.2 Chitosan and glucosamine

5.2.1 Introduction

Shellfish and crustacean shells are rich in chitin, which is the raw material of chitosan and N-acetyl-glucosamine. It is suggested that chitosan and its oligomers act as fat scavengers in the digestive tract and remove fat and cholesterol via excretion (Kim and Mendis 2006). Based on both *in vitro* and *in vivo* studies, it has also been reported that chitosan oligomers can possess anti-tumour activities due to the immunostimulating effects of chitin and its carboxymethyl derivatives via stimulation of cytolytic t-lymphocytes (Kim and Mendis 2006). A wound-healing effect of chitin and chitosan has further been suggested which was found to be associated with their immunostimulating property or by their ability to affect the fibroblast growth factor (Kim and Mendis 2006). Chitosan is also a versatile polymer used for food preservation and drug delivery. According to Ruel-Gariepy *et al.* (2006), it has the ability to improve cartilage repair in a safe manner (Ruel-Gariepy *et al.* 2006).

5.2.2 Role of chitosan in regulating cardiovascular risk factors.

As reviewed by Muzzarelli (1999), dietary chitosan was found to affect serum cholesterol and atherosclerosis in normal and diabetic mice. The review also described that in humans, it could be used to treat hypercholesterolemia, and also that it exhibits anti-ulcer, anti-arthritic, anti-hypertensive and anti-uricaemic properties. Based on the statistical analysis of a series of human studies, the same review also indicates that chitosan might be useful to control overweight when associated with a diet.

5.2.3 Antibacterial properties of chitosan.

Chitosan has been shown to demonstrate antibacterial activities on a wide variety of bacteria, and bacterial cell membrane damage is supposed to be the active principle (Liu *et al.* 2004a). A recent study, however, indicated that the decreased pH value caused by organic acids in the chitosan solution added was responsible for the antibacterial activity of chitosan solution against *L. pneumophila* and *E. coli* (Fujimoto *et al.* 2006).

5.2.4 Cancer

In vivo experiments were done with mice bearing sarcoma[40] 180 solid or uterine cervix[41] carcinoma tumor cells. Consumption of high molecular weight chitosan oligosaccharides (1.5-5.5 kDa), 20 mg/kg/day or more, improved survival and number of survivors after 40 days of feeding (Jeon and Kim 2002).

5.2.5 Immuno-stimulation of macrophage function.

Oligochitosan containing 3-10 monomer residues was prepared. *In vitro* macrophage activation was studied in a mice murine macrophage cell line. The mannose-receptor (MR) was found to be the major receptor responsible for oligochitosan uptake. MR was also engaged in oligochitosan induced TNF-α enhancement (Han *et al.* 2005). The effects of various chitin/chitosan compounds on the migrations of mouse peritoneal macrophage[42] (PEM) and rat macrophage cell line (rMp) were evaluated *in vitro*. The migratory activity of PEM was enhanced significantly by chitin and chitosan oligomers, but reduced by chitin, chitosan and glucosamine. The migratory activity of rMp was increased significantly by chitin and chitosan polymers and by glucosamine (Okamoto *et al.* 2003).

5.2.6 Wound healing

The breakdown of glycosaminoglycans causes inflammation at mucosal surfaces. It has been reported that N-acetyl glucosamine can be used as a substrate for tissue repair mechanisms. In a pilot study of N-acetyl glucosamine, children with inflammatory bowel diseases (Crohn's disease or ulcerative colitis) were orally administered N-acetyl glucosamine, and 8 out of 12 children showed a clear improvement (Salvatore *et al.* 2000). In an *in vitro* study, the effects of chitin and chitosan on human dermal fibroblast[43]-mediated contraction of collagen lattices[44] were examined to model the contraction of cutaneous wounds as it might appear *in vivo*. The results indicated that highly deacetylated chitosan inhibits fibroblast-mediated contraction of collagen lattices and may therefore be useful as a therapeutic agent to reduce contraction during wound healing (Howling 2002). According to Schaafsma (2007) there is, however, no sufficient evidence for retardation or prevention of osteoarthritis by glucosamine.

---

[40] A sarcoma is a cancer of the connective or supportive tissue (bone, cartilage, fat, muscle, blood vessels) and soft tissue.
[41] The cervix is the lower, narrow portion of the uterus where it joins with the top end of the vagina.
[42] Macrophages are cells within the tissues that originate from specific white blood cells called monocytes.
[43] A fibroblast is a type of cell that synthesises and maintains the extracellular matrix of many animal tissues.
[44] An *in vitro* dermal equivalent that has led to the development of an original model of dermal tissue. A collagen lattice as a cell culture matrix allows the growing of cells in 3 dimensions.

5.2.7 Protection against Alzheimer's disease

ß-secretase is an aspartic protease present at high levels in neurons of the brain, and the activity of this enzyme is related to the development of Alzheimer's disease. Hence, partial inhibition may reduce the risk of adapting this disease. Nine kinds of chito-oligosaccharides with different degrees of deacetylation were prepared from crab shells. In *in vitro* experiments, non-competitive inhibition of ß-secretase was obtained with 90% deacetylated chito-oligo-saccharides with a molecular weight of between 3 and 5 kDa (Byun and other 2005). The microtubule[45]-associated human brain protein, tau[46], abnormally hyperphosphorylated and aggregated into neurofibrillary tangles in brains of individuals with Alzheimer's disease (AD). *In vitro* experiments showed that such abnormal hyperphosphorylation was inhibited by N-acetyl-glucosamine. *In vivo* experiments with mice indicated that deficient brain glucose uptake may induce AD (Liu *et al.* 2004b).

Table 9. *Overview of reviews, experimental studies, animal studies and human intervention studies related to biological activities from chitosan and glucosamine cited in the present review.*

End point	Reviews	Interventions	Animal studies	Experimental studies
High cholesterol	Muzzarelli 1999; Kim and Mendis 2006			
Infection	Ruel-Gariepy *et al.* 2006			Liu *et al.* 2004a; Fujimoto *et al.* 2006
Cancer	Kim and Mendis 2006		Jeon and Kim 2002	
Low immune response	Kim and Mendis 2006			Okamoto *et al.* 2003; Han *et al.* 2005
Wound	Kim and Mendis 2006; Ruel-Gariepy *et al.* 2006; Schaafsma 2007	Salvatore *et al.* 2000		Howling 2002
Alzheimer's disease	Kim and Mendis 2006		Liu *et al.* 2004b	Byun *et al.* 2005

[45] Microtubules are one of the components of the cytoskeleton being a cellular skelleton contained within the cytoplasm.

[46] Tau in biochemistry is a protein associated with microtubules.

5.3 Chondroitin sulphate

5.3.1 Introduction

Chondroitin sulphate is a glycoaminoglycan; a sulphated polysaccharide mainly present in cartilage tissue. It is used as a component in dietary supplement or health food for the treatment of osteoarthritis[47] (OA). It is suggested that it not only has a pain-relieving effect but also stops the progressive destruction of articular cartilage and even rebuilds it (Goggs *et al.* 2005; Bruyere and Reginster, 2007). However, Schaafsma (2007) points out that recent studies suggest that chondroitin sulphate mainly has anti-inflammatory properties that must be due to the sulphate part of the molecule. Furthermore, chondroitin sulphate is believed to be harmless with no side-effects in contrast to the more commonly used nonsteroidal anti-inflammatory drugs (NSAID). Commercial products are mainly produced from bovine and pig cartilage and to some extent from cartilaginous fishes (sharks/stingrays). Utilisation of chondroitin sulphate from cartilage from bony fishes is probably limited due to the relatively small amount of cartilage in bony fish. However, the substantial increase in aquaculture is resulting in large amounts of homogenous by-products such as heads which contain the main part of cartilage in bony fish and could therefore be a new and beneficial raw material source of these ingredients. Furthermore, there is limited knowledge of biochemical characterisation of bony fish chondroitin sulphate and whether it differs in composition and functionality from mammal chondroitin sulphate. The literature referred to here regarding the health effects of chondroitin sulphate is therefore primarily based on studies on chondroitin sulphate from other sources.

5.3.2 Evidence regarding the treatment of osteoarthritis

A large number of randomised control trials (RCT's) studying the efficacy of glucosamine/chondroitin sulphate on OA, have been carried out during the last forty years. A review from 2006 of four large meta-analyses of RCT's carried out between 1966 and 2005 concluded that results on the efficacy of glucosamine and chondroitin sulphate in treatment of OA are still inconclusive mainly due to the weak research design of many of the studies (Distler and Anguelouch 2006). This is consistent with the meta-analysis carried out by McAlindon *et al.* (2000) who reviewed 37 trials carried out between 1980 and 1998. Of 37 studies, they only found 15 studies that passed the quality criteria that in their view resulted in valid data. Of the 15 studies, 9 studies used chondroitin sulphate. Their main conclusion was that there was an effect of chondroitin sulphate on pain in connection with osteoarthritis. However, this effect was diminished when only high-quality and large trials were considered,

[47] Degenerative joint disease: a low grade inflammation in the joints resulting in pain.

and they furthermore concluded that larger and private company independent studies were necessary to verify the efficacy of chondroitin sulphate.

Results from another meta-analysis on the efficacy of chondroitin sulphate alone based on 7 RCT´s (Leeb *et al.* 2000) between 1991 and 1998 also concluded that chondroitin sulphate may reduce pain in treatment of OA. A more recent meta-analysis carried out in 2003 (Richy *et al.* 2003) on 15 placebo-controlled RCT´s between 1980 and 2002 concluded that chondroitin sulphate was more efficient than placebo in pain treatment of OA. However, both studies also concluded that longer and larger trials are needed to confirm the positive effect of chondroitin sulphate.

A very large randomised trial was initiated in 2000 by the National Institute of Health, USA, with 1588 patients who were treated with glucosamine or chondroitin sulphate or a combination for 24 weeks (Clegg *et al.* 2006). The results showed that chondroitin sulphate, alone or in combination with glucosamine, did not reduce pain effectively in the overall group of patients with OA. However, the results indicated that a combination of chondroitin sulphate and glucosamine was effective in a subgroup of patients with moderate to severe knee pain. In spite of the size of the trial, it was recognised that it also had a number of limitations, among them a high number of patients with low knee pain and a very high placebo response, making it difficult to give firm conclusions.

However, a very recent review (Bruyere and Reginster 2007), evaluating the results of a number of recent randomised controlled trials and meta analyses, including the above, concludes that the overall result suggests that chondroitin sulphate has a small to moderate effect on pain in the treatment of OA and it also seems that it has a positive structure-modifying effect in the knee. However, other studies question whether it has a cartilage-repairing effect (Schaafsma, 2007).

5.3.3 Other potential effects

There may be other perspectives for using chondroitin sulphate from bony fish as health food or pharmaceuticals. A study showed that chondroitin sulphate isolated from salmon nasal cartilage inhibited fat absorption in mice and therefore could be a possible agent to prevent obesity in connection with high-fat diets (Han *et al.* 2000).

Recent studies show that chondroitin sulphate is not only a structural component in cartilage but also encodes information by specific sulphation pattern and can bind growth factors, neurotrophic factors[48] and affect neuronal growth[49]. Thus, it may

[48] Proteins that help nerve cells to survive.
[49] Growth of nerve cells.

have the potential as a therapeutic that can control the activity of growth factors for cancer or neuroregeneration[50] (Gama and Hsieh-Wilson 2005). A Japanese research group (Nandini *et al.* 2005) has isolated a novel chondroitin sulphate/dermatan sulphate chain from shark skin, which exhibited neurotrophic activity and binding activities for growth factors *in vitro*. They concluded that it had potential as an ingredient in therapeutics.

Results from recent studies carried out on rats also show that chondroitin sulphate has free radical-scavenging properties *in vitro* suggesting a potential use in health food (Nandini *et al.* 2005; Xiong and Jin 2007).

Table 10. *Overview of reviews, observational studies, intervention studies, animal studies and in vitro studies on chondroitin sulphate and health cited in the present review.*

End point	Reviews	Observational studies	Interventions	Animal studies	Experimental studies
Osteoarthritis	Goggs *et al.* 2005; Distler and Anguelouch 2006; Bruyere and Reginster 2007; Schaafsma 2007	Leeb *et al.* 2000; McAlindon *et al.* 2000; Richy *et al.* 2003	Clegg *et al.* 2006		
Obesity/weight loss				Han *et al.* 2000	
Cancer	Gama and Hsieh-Wilson 2005				
Oxidative stress				Xiong and Jin 2007; Xiong *et al.* 2007	
Neuro-related diseases	Gama and Hsieh-Wilson 2005				Nandini *et al.* 2005

[50] Regeneration of nerve cell.

5.4 LC n-3 PUFA fortification

5.4.1 LC n-3 PUFA fortification of food

According to section 3.1, there is considerable evidence that regular LC n-3 PUFA consumption is associated with significant health benefits. Western societies consume very little fish and fish products, and thus usually only small amounts of LC n-3 PUFA. Increasing consumption to recommended levels (Kris-Etherton *et al.* 2002) would involve major alterations in dietary habits. An alternative strategy for increasing LC n-3 PUFA intake may be to provide a wide range of commercial food products and ingredients fortified with fish oil, which can be incorporated into an existing diet (Metcalf *et al.* 2003).

Providing human subjects with everyday food items fortified with fish oil has been shown to increase the intake of LC n-3 PUFA (Lovegrove *et al.* 1997; Mantzioris *et al.* 2000; Metcalf *et al.* 2003). In the study by Metcalf *et al.*, subjects were allowed to incorporate fortified food into their domestic food preparation in accordance with their own tastes and preferences. LC n-3 PUFA intakes of about 1 g/day were readily achievable. Similar investigations by Lovegrove *et al.* (1997) and Mantzioris *et al.* (2000) achieved even higher intakes of EPA+DHA up to 1.8 g/day. When participants were allowed to choose from a variety of fortified products, enriched spread, milk and sausage were the most frequently eaten fortified food items (Metcalf *et al.* 2003).

The bioavailability of LC n-3 PUFA in enriched food has been demonstrated by an increase of the DHA and EPA in human erythrocyte membrane PL in several studies (Saldeen *et al.* 1998; Visioli *et al.* 2000; Wallace *et al.* 2000). Even minimal enrichment which provided a daily LC n-3 PUFA intake of 300 mg in humans increased EPA and DHA in plasma PL by 32 and 18%, respectively (Saldeen *et al.* 1998).

Concerns have been raised about the oxidative stress that can result from increased intakes of LC n-3 PUFA (Kris-Etherton *et al.* 2000; Sanders 2000). It has been reported that fish oil increases lipid peroxidation markers in patients with primary hypertriglyceridaemia (Stalenhoef *et al.* 2000) but there are several other human studies which do not show increased lipid peroxidation in plasma or LDL after intake of fish oils (Higdon *et al.* 2000, 2001; Higgins *et al.* 2001 reviewed in Robinson and Stone 2006). Although it is not always clear whether oxidation products are formed in the food during storage or food preparation, or in the body by physiological or pathological processes, it is clear that enriched products require precise oxygen elimination from the packaging, and stabilisation with appropriate levels of antioxidants addition in order to deliver high quality and safe enriched food to the costumers (Kolanowski *et al.* 1999).

It is a promising strategy to offer enriched foods alongside ordinary products on the shelves to increase LC n-3 PUFA intake. However, the population would have to be made aware of the benefits of actively choosing these specific products, and of most probably paying a higher price for them (Engstrom *et al.* 2003).

5.4.2 Infant formula[51] fortification with LC n-3 PUFA

Formula-fed infants have lower levels of DHA in their brain, plasma, and erythrocyte PL layers[52] than their breastfed counterparts (Ponder *et al.* 1992; Makrides *et al.* 1994). The question is whether adding LC n-3 PUFA to formula produces improved infant development. Two recent Cochrane reviews (Simmer and Patole 2004; Simmer 2004) concluded that available evidence does not support infant development benefits from LC n-3 PUFA supplementation. Results were largely based on randomised controlled intervention trials (reviewed in Lewin *et al.* 2005) and indicated that neither growth patterns nor visual function were affected by the intake of supplemented LC n-3 PUFA in either term or preterm infants[53]. The neurological development outcomes were influenced somewhat in preterm infants although not all of the studies found evidence for a benefit. In the preterm population, the only type of clinical outcome that showed a significant favourable effect related to the intake of LC n-3 PUFA was the Fagan test of Infant Intelligence at 6 and 12 months of age. This test assesses cognitive function. However, scores on other developmental scales, e.g. Bayley's Developmental Index (MDI) were not influenced by infant supplementation at any age.

Three more recent intervention studies (Clandinin *et al.* 2005; Fang *et al.* 2005; Lewin *et al.* 2005) showed positive effects of LC n-3 PUFA supplements in preterm infants, but evidence is as yet inconclusive (reviewed in Eilander *et al.* 2007). Taken together, there are indications that LC n-3 PUFA supplementation has a beneficial effect on the cognitive development of preterm infants. It may be that the effects of LC n-3 PUFA supplementation are greater when the duration of supplementation is continued until the age of 12 months after term; or that the effect can be better detected at older ages, when cognitive tests are more sensitive and reliable.

Most of the more recent studies on LC n-3 PUFA supplementation in term infants[54] failed to demonstrate the effects on various aspects of cognitive development (reviewed in Eilander *et al.* 2007). Future studies are needed to assess the effect of LC n-3 PUFA supplementation on cognitive development at high doses (100 mg DHA plus 200 mg AA) and prolonged duration (preferably 12 months).

---

[51] Milk substitute for babies.
[52] Outer layer (skin) of red blood cells.
[53] Baby born too early.
[54] Born after normal length pregnancy.

Table 11. *Overview of reviews and human intervention studies on the effects of LC n-3 PUFA fortified infant formulas/food cited in the present review.*

Subjects/End point	Reviews	Interventions
Preterm infants	Simmer and Patole 2004; Lewin *et al.* 2005; Eilander *et al.* 2007	Fewtrell *et al.* 2004; Clandinin *et al.* 2005; Fang *et al.* 2005
Term infants	Simmer 2004; Lewin *et al.* 2005; Eilander *et al.* 2007	
Fortification increases intake		Metcalf *et al.* 2003; Lovegrove *et al.* 1997; Mantzioris *et al.* 2000
Lipid peroxidation	Robinson and Stone 2006	Stalenhoef *et al.* 2000; Higdon *et al.* 2000, 2001

6. Potential risks associated with fish consumption – what are they?

6.1 Introduction

Recommendations for fish intake are similar in the Nordic countries to other European and North American food agencies. There are both general recommendations (fish at least two times/week) and detailed recommendations for limited intake of specific species to avoid high exposure to methylmercury and dioxins. For pregnant women, women of childbearing age and high fish consumers these limitations are stricter. There are also prescribed maximum levels for dioxins and methylmercury in seafood allowed on the European Union market.

The main issues when it comes to contaminants in seafood are methylmercury, dioxins and dioxin like PCBs. The concentrations of these contaminants vary largely between fishing waters and fish species.

6.2 Methylmercury

Mercury is converted to methylmercury by microbial activity in seas and lakes. Unlike elemental and inorganic mercury, methylmercury is both absorbed and transported to tissues. It is lipid soluble and bioaccumulates in aquatic food chains, thus long living predatory fish could contain high concentrations of methylmercury.

Very high mercury exposures from industrial accidents are well documented and cause paresthesias, ataxia and sensory abnormalities in adults and delayed cognitive and neuromuscular development in children exposed during pregnancy. The health risks with chronic low exposure of methylmercury are not as well documented

(Mozzafarian and Rimm 2006). Studies on children in several countries have shown a relationship between higher gestational maternal exposure to mercury and lower scores on some neurological tests but not others. In other studies, gestational maternal fish intake was associated with improved neurodevelopmental scores, probably due to the LC n-3 PUFA from the fish (Mozzafarian and Rimm 2006). High mercury levels in the body may increase the risk of CVD, but these findings needs to be confirmed (Virtanen *et al.* 2007). Investigations of the effects of mercury exposure from high fish consumption show conflicting results concerning CVD. However, the net effect of fish consumption tends to be positive. Based on numerous studies of consumption of fish with various methylmercury levels, positive net effects on cognitive function in elderly adults have also been reported (Mozzafarian and Rimm 2006).

There is no general common geographic trend for mercury in seafood. The largest problem with methylmercury is in fresh water fish and in large predatory fish species in seawater.

FAO/WHO recommends an allowable upper limit of weekly intake of methylmercury of 1.6 μg/kg body weight based on the most sensitive toxicological end-point (developmental neurotoxicity). The U.S. Food and Drug Administration (FDA) and Environmental Protection Agency (EPA) reference dose is significantly lower; 0.1 μg/kg body weight per day. The upper limit for methylmercury in fishery products is 0.5 μg/g wet weight in EU (Commission Regulations 2006).

6.3 Persistent organic pollutants (POPs): dioxins and polychlorinated biphenyls (PCBs)

Meat, dairy products and fish are the largest sources of POP exposure. POPs have shown to be carcinogenic in animal experiments and there are some indications of this also in humans. They may also have negative effects on the immune system, central nervous system, hormone levels and reproduction. The only persistent effect associated with dioxin exposure in humans is chloracne. Prenatal exposure of PCB and dioxins has been associated with neurodevelopmental deficits in children, but there are contradicting reports.

Polychlorinated dioxins (PCDD) and dibenzofurans (PCDF) are usually called dioxins. To facilitate the toxicity assessment of dioxins and dioxin like PCBs, toxic equivalences (TEQ) are used. In EU, the maximum allowed total dioxin and dioxin like PCBs (WHO-PCDD/F-PCB-TEQ) in fish and fish products on the market are set to 8 pg TEQ/g wet weight and out of these, maximum 4 pg PCDD/F-TEQ/g wet weight of dioxin (Commission Regulation 2006). The Scientific committee for food in the EU recommends a total daily intake of dioxin and dioxin-like PCBs up to a maximum 2 pg TEQ/kg body weight (Interim Report 5 2002).

6.4 Risks and benefits of seafood

A few review studies have tried to weigh the negative effects of methylmercury and the positive effects of LC n-3 PUFA from fish against each other. Gochfeld and Burger (2005) present the results (based on CVD and prenatal development) in a benefit-risk dose curve. To gain health benefits from fish, 15 g/day was needed and for a maximum health benefit in CVD the threshold was set to 45 g fish/day. For fish containing 0.1 µg/g of methylmercury the negative threshold was set to 60 g fish/ day and for fish containing 0.23 µg/g of methylmercury the threshold was set to 30 g fish/day. The risk-benefit of changes in fish intake of the whole population was investigated in another study. They concluded that if women of childbearing age substitute fish high in methylmercury to fish low in methylmercury it would yield developmental benefits, but if they just decrease their fish intake the benefits are reduced. If other adults decrease their fish intake, the net public health impact is negative (Cohen *et al.* 2005). Common conclusions are to minimise consumption of highly contaminated fish species, but otherwise to consume fish 1-2 times a week to achieve the beneficial effects (Costa 2007).

As fish not only contains methylmercury, but also POPs and other contaminants with different negative or unspecified negative health effects, it is hard to evaluate the total risk of fish intake. This is especially so because of the large differences that exist with respect to vulnerability of the populations exposed to these pollutants. Not forgetting that the content of both contaminants and nutritional composition (like EPA and DHA) varies between species, between fishing waters and even between seasons and it is thus impossible to draw general conclusions for the risk/benefit related to fish consumption.

7. Conclusions and future recommendations

This review illustrates that there is strong evidence showing that the consumption of whole fish and fish-derived LC n-3 PUFA can positively affect CHD and CHD mortality, among other things. Some evidence, although mainly from animal and *in vitro* studies, also exists to strengthen the notion that LC n-3 PUFA are not the only carriers of these effects. Unfortunately, many of the studies of 'non-n-3 compounds' show contradicting results, which prevent a clear consensus on their role in human health. Besides differences in the study designs used, an important source of difference most likely lies in the specific nutrigenomics of the subjects studied. Blood cells behave very differently in different individuals when they are exposed to agonists. In the same way, the response to different nutrients is quite different and distinct for each individual. Thus, the biological variety makes the nutrient very good for one person but less effective in another.

In conclusion, the findings on **fish** and cardiovascular health show that low consumption of fish (1-3 times/month) compared to no fish consumption decreases the risk of CHD and CHD-mortality. Increasing fish consumption further decreases the risk. Also, the evidence is stronger in secondary prevention. Fish intake is also associated with a decreased risk of stroke, arrhythmia and atherosclerosis. Fish consumption has been shown to increase bleeding time and HDL levels, to reduce blood TAG levels (mainly fatty fish), while it has minor effects on blood pressure. No firm conclusions can be drawn from the few studies carried out to study how fish intake affects oxidative stress, inflammation, diabetes, weight, and appetite regulation.

Regarding fish intake and cancer, most studies performed are prospective studies and case-control studies. These have shown either no association or reduced risk at high fish intake. Digestive tract cancer is most evaluated and most evidence pointing in a positive direction has been found here. In general, there are too few studies and not enough quality to draw any firm conclusions on fish intake and cancer.

When it comes to fish and inflammatory/immune diseases, several studies show relationship between fish consumption during pregnancy and decreased risk for early childhood atopy-related outcomes (asthma, allergy, etc.). Furthermore, fatty fish could influence the onset of rheumatoid arthritis.

Regarding brain/cognition, it has been found that fatty fish consumption is associated with less cognitive decline. Generally, fish intake is also associated with a lower incidence of depression - especially for women. It can also be said that fish intake during pregnancy and infant fish intake influence cognitive development positively.

From a future perspective, more intervention studies regarding fish and cardiovascular health are needed that are more specific when it comes to fish species (e.g. fatty/lean) and type of cooking method used. Also, new markers for CVD must be followed such as CRP and ox-LDL. In the field of cancer, more, and better controlled studies on specific cancer types are needed. In the areas of obesity, inflammatory/immunological diseases, cognitive decline and development and mood disorders, intervention studies are needed.

Based on reviews from the last five years, there is strong evidence to support the beneficial effects of the fish-derived **LC n-3 PUFA** on CVD outcomes. There is also strong evidence that LC n-3 PUFA can lower TAG levels. Furthermore, there is good evidence that the LC n-3 PUFA improve visual development in preterm infants and improve clinical outcome of early-stage schizophrenia. Some evidence also exists for beneficial effects of LC n-3 PUFA on unipolar depression.

Based on current knowledge, future studies should systematically determine which are the most effective types of n-3 PUFA, and also which is the best regimen (dose and time).

One lipid class that has lately gained a lot of attention are the **phospholipids (PL).** In conclusion, some PL are reported to have a positive effect on cardiovascular health and brain. The PL group called sphingolipids is reported to have a role in the suppression of colon cancer. Most studies so far have been performed on vegetable and animal PL and the mechanisms are not well understood. The marine PL are also associated with positive health effects due to their high LC n-3 PUFA content. Some studies have also shown that the bioavailability of the LC n-3 PUFA is higher from PL than from TAG.

From a future perspective, an increase in human clinical studies on the health effects of the marine PL is expected due to the increased focus on utilisation of PL-containing fish by-products and the increased production of synthetic PL with marine LC n-3 PUFA incorporated.

Fatty fish is a very important dietary source of **vitamin D**. This is of particular value for the general Nordic population during wintertime, since there is not enough sunlight to produce vitamin D in the skin. Vitamin D has an important role in bone mineralisation and deficiency causes rickets in infants and children and osteomalacia in adults. Observational studies have also found an inverse relationship between vitamin D status and vascular diseases as well as some risk factors for CVD. Epidemiological studies have shown that vitamin D status is inversely related to several forms of cancer; in particular colorectal cancer. Finally, adequate vitamin D concentrations are important during pregnancy, e.g. to ensure appropriate maternal responses to the calcium demands of the foetus and neonatal handling of calcium. Deficiency has also been proposed to be a risk factor for maternal pre-eclampsia.

In the future, the associations found between vitamin D insufficiency and vascular diseases, diabetes, cancer, brain function, pregnancy and rheumatoid arthritis must be confirmed in long-term intervention studies.

Within the field of **seafood-derived proteins**, peptides and amino acids, all being strong candidates as carriers of health effects that goes beyond LC n-3 PUFA, the following conclusions can be drawn. High-protein diets in general have been shown to help with weight control. Regarding the specific effects of fish proteins, one human study has shown that cod proteins, but not other animal proteins, improved insulin sensitivity in insulin resistant individuals. This could contribute to the prevention of type II diabetes. Using animal experimentation, feeding defatted cod protein to diabetic or hypertensive rats has shown positive effects on insulin sensitivity and high blood pressure, but the underlying mechanism remains obscure.

In future research, both animal models and human clinical trials are needed to permit identification of the exact bioactive components of fish protein. A special challenge here is to perform studies with completely lipid-free fish proteins so as not to confuse the results with the effects, for example, of LC n-3 PUFA. Also, to further understand what is unique about fish proteins, there is a need for studies where other animal muscle proteins are used as controls.

Peptides occur as a result of *in vivo* synthesis or by random hydrolysis of proteins using endogenous or exogenous proteolytic enzymes. A great number of small- and medium-sized marine peptides may provide valuable functions like stimulation of non-specific immuno-defence systems, reduction of high blood pressure as well as protection against anaemia, harmful oxidation and cancer development. However, so far, most of these observations have been obtained by animal or *in vitro* experiments.

In the future, there is a great demand for human intervention studies to verify whether results obtained by animal or *in vitro* experiments may provide useful medical benefits. As for proteins, there is a special need to evaluate hydrolysates completely free of lipids, and to compare marine hydrolysates with hydrolysates from other muscle sources. Identification of the specific peptides that affect various end-points is also desired.

A reduced CVD risk from the intake of the free amino acid **taurine**, alone or in combination with LC n-3 PUFA in seafood, has been suggested from both human and animal trials. Among the reasons given are an antioxidant activity, a blood pressure lowering effect, an improvement of cardiac performance, a suppressed platelet aggregation, a cholesterol lowering effect and the capacity to suppress the development of atherosclerotic lesions. Recently it was shown that simultaneous intake of LC n-3 PUFA and taurine by healthy subjects had additional beneficial effects over intake of LC n-3 PUFA alone, in particular upon blood lipids.

As the research in the area of taurine and other amino acids is scarce, at least when compared to the LC n-3 PUFA, it should allow for many hypotheses to be explored. Further studies, in animal and human models, are warranted to allow for the calculation of the separate effects of taurine and other amino acids, dose dependence and last but not least the combined effects of components available from a seafood diet.

Seafood is from a nutritional point of view an important source of **selenium**. A higher consumption of seafood, and increasing the Se content in farmed fish may contribute to an improved Se status in humans. Limited results have shown a large variation in Se-bioavailability. There are strong indications that Se has a cancer prevention effect when a supranutritional dose is given. The speciation of

Se (γ-glutamyl-SeMeSeCys and SeMeSeCys) also seems to be important in that respect.

In the future, studies of the bioavailability of Se from farmed fish using fish enriched with organic selenium via dietary modulation might give new insights into the absorption and retention of Se in humans. Using 'Se tailor-made farmed fish' will provide the opportunity to control the Se feed source, the speciation and the level of selenium in order to minimise variation in bioavailability due to the seafood source. Using 'Se tailor-made farmed fish' might also be a good way to study the possible anti-carcinogenic effects of Se in animal models.

In 2001, discards from the world's fisheries exceeded 20 million tons, equivalent to 25% of the total production of marine capture fisheries. If we succeed to utilize more **seafood by-products** as human food and/or as ingredients in foodstuff, health foods, nutraceuticals, pharmacy, cosmetics etc., it has been estimated that the value adding may increase up to 5-fold. Examples of components that can be extracted from by-products are chitosan, glucosamine and chondroitin sulphate. In addition to value adding, the use of marine components as ingredients Is also a way to make the benefits of fish consumption available to large numbers of people who have limited opportunities to (or do not chose to) include fish as part of their normal diet. Among the latter are infants, people with fish allergies and vegetarians. To date, the most common marine ingredients by far are the LC n-3 PUFA.

Chitosan can be produced from shellfish and crustaceans shells. Potential health effects of chitosan, chitooligomers and (N-acetyl-) glucosamine have been shown in the scientific literature; mainly from animal and *in vitro* studies. Among these described effects are antibacterial effects, antioxidative activities, and activities as fat scavengers in the digestive tract. There are indications that chitosan oligomers have immunomodulating activity, can enhance wound healing, and increase the survival of mice bearing carcinoma tumour cells. Moreover, highly deacetylated oligomers of chitosan were shown to partially inhibit β-secretase and may prevent the development of Alzheimer's disease. Human studies have shown that chitosan can be used to treat hypercholesterolemia and to control overweight.

Significantly more documentation on human health effects will be necessary to increase the use of chitosan and its derivatives in the high priced markets. A substantial amount of work and collaboration between scientists with their background both in medical and technical areas will here be required.

Cartilage from bony fishes can be used for isolation of **chondroitin sulphate**. This compound has mainly been studied in relation to osteoarthritis. However, as effects of chondroitin sulphate from bony fishes on osteoarthritis have not been studied specifically in RCT´s, there is no knowledge whether marine chondroitin sulphate is more efficient e.g. than mammalian chondroitin sulphate. Chemical/structural differences depending on raw material source are also unknown, as well as potential differences in biological activity when it comes to inhibition of fat absorption, binding of growth/neurotrophic factors and antioxidative properties. However bony fish can be advantageous as raw material for chondroitin sulphate in countries where religious reasons makes it more attractive to use marine based chondroitin sulphate.

In a future perspective, there is a need for basic research on characterisation of chondroitin sulphate from bony fish species regarding chemical, physical and structural properties and also research that can evaluate whether bony fish chondroitin sulphate have different biological activities compared to mammalian chondroitin sulphate. There is only a relative small amount of chondroitin sulphate in bony fish, mainly located in the nose part of the head where it makes a total of 0.3%. Research on how the whole raw material - the head - can be fully utilised by transforming protein/lipid fraction into high-value products is therefore necessary.

Currently, the evidence for beneficial effects of **LC n-3 PUFA supplementation of infant formula** on infants' development is weak and inconsistent. However, this is most likely explained by different study designs and interfering background diets. As indicated by recent promising intervention studies, future research should focus on adequate doses (e.g. 100 mg DHA plus 200 mg AA), longer periods of supplementation (up to 12 months) and follow up (preschool and school age) in order to assess possible benefits of LC n-3 PUFA supplementation of infant formula.

With regards to **LC n-3 PUFA fortification of food**, there is no doubt that fortification can improve health of the consumers. Future research has to address issues of stability, shelf life and lipid oxidation of fortified food in order to guarantee best possible olfactory results and health benefits.

According to above, there are numerous interesting components present in both the edible and non-edible parts of seafood that according to different mechanisms can be beneficial to health. In this context, we have also chosen to highlight some potential **contaminants** in seafood that could be hazardous. Although fewer and much more rare, these contaminants are often highlighted in media, and thus become of more concern to public than the beneficial compounds. The main contaminants in seafood are methylmercury, dioxins and dioxin like PCBs. However, it must be stressed that the concentrations of these contaminants largely vary between fishing waters and fish species. Only few species and geographic locations are considered a risk.

High **mercury** levels in the body may increase the risk of CVD, but these findings needs to be confirmed. Investigations on effects from mercury exposure from high fish consumption on CVD and cognitive function have been carried out, but point at the net positive effect of fish consumption. A common conclusion is thus to minimize consumption of highly methyl mercury contaminated fish species, but otherwise to consume fish 1-2 times a week to achieve the beneficial effects.

Dioxins and PCB have been shown to be carcinogenic in animal trials, with some indications also in humans. They may also cause chloracne, have negative effects on the immune system, central nervous system, hormone levels and reproduction. Within EU, a maximum total daily intake of dioxin and dioxin like PCBs to maximum 2 pg TEQ/kg body weight.

Because of the risk that fear for contaminants keeps the public away from eating fish, several **risk-benefit studies** of fish have been carried out. One such study concluded that it would be beneficial if women of childbearing age substitute fish with high methylmercury levels to low methylmercury fish, but if they just decrease their fish intake the benefits are reduced. If other adults decrease their fish intake, the net public health impact is negative. The latter is given as one important take-home message from the present review.

References

Abe H. 2000. Role of histidine-related compounds as intracellular proton buffering constituents in vertebrate muscle. Biochemistry-Moscow 65(7):757-765.

Aggett PJ, Antoine JM, Asp NG, Bellisle F, Contor L, Cummings JH, Howlett J, Müller DJ, Persin C, Pijls LT, Rechkemmer G, Tuijtelaars S, Verhagen H. 2005. PASSCLAIM: consensus on criteria. Eur J Nutr 44(Suppl 1):i5-30.

Aid S, Vancassel S, Linard A, Lavialle M. Guesnet P. 2005. Dietary docosahexaenoic acid [22:6(n-3)] as a phospholipid or a triglyceride enhances the potassium chloride –evoked release of acetylcholine in rat hippocampus. Journal of Nutrition 135:1008-1013.

Ait Yahia D, Madani S, Prost J, Bouchenak M, Belleville J. 2005. Fish protein improves blood pressure but alters HDL2 and HDL3 composition and tissue lipoprotein lipase activities in spontaneously hypertensive rats. Eur J Nutr 44(1):10-7.

Ait-Yahia D, Madani S, Savelli JL, Prost J, Bouchenak M, Belleville J. 2003. Dietary fish protein lowers blood pressure and alters tissue polyunsaturated fatty acid composition in spontaneously hypertensive rats. Nutrition 19(4):342-346.

Åkesson B, Skrikumar TS. 1994. Occurrence of low-molecular-weight and high-molecular-weight selenium compounds in fish. Food Chemistry 51:45-59.

Ali MM, Vaidya V. 2007. Vitamin D and cancer. J Cancer Res Ther 3(4):225-30.

Amate L, Gil A, Ramirez M. 2001a. Feeding Infant Piglets Formula with Long-Chain Polyunsaturated Fatty Acids as Triacylglycerols or Phospholipids Influences the Distribution of These Fatty Acids in Plasma Lipoprotein Fractions. Journal of Nutrition 131(4):1250-1255.

Amate L, Gil A, Ramirez M. 2001b. Dietary Long-Chain Polyunsaturated Fatty Acids from Different Sources Affect Fat and Fatty Acid Excretions in Rats. Journal of Nutrition 131(12):3216-3221.

Appleton KM, Hayward RC, Gunnell D, Peters TJ, Rogers PJ, Kessler D, Ness AR. 2006. Effects of n-3 long-chain polyunsaturated fatty acids on depressed mood: systematic review of published trials. Am J Clin Nutr 84(6):1308-16.

Balk EM, Lichtenstein A, Chung M, Kupelnick B, Chew P, Lau J. 2006a. Effects of omega-3 fatty acids on coronary restenosis, intima-media thickness, and exercise tolerance: a systematic review. Atherosclerosis 184(2):237-46.

Balkan J. Öztezcani S, Hatipogluz A, Çevikbas U, Aykac-Toker G, Uysal M. 2004. Effect of a taurine treatment on the regression of existing atherosclerotic lesions in rabbits fed on a high-cholesterol diet. Bioscience Biotechnology and Biochemistry 68(5):1035-1039.

Balk EM, Lichtenstein A, Chung M, Kupelnick B, Chew P, Lau J. 2006b. Effects of omega-3 fatty acids on serum markers of cardiovascular disease risk: a systematic review. Atherosclerosis 189(1):19-30.

Bao DQ, Mori TA, Burke V, Puddey IB, Beilin LJ. 1998. Effects of dietary fish and weight reduction on ambulatory blood pressure in overweight hypertensives. Hypertension 32:710-717.

Barberger-Gateau P, Jutand MA, Letenneur L, Larrieu S, Tavernier B, Berr C 2005. Correlates of regular fish consumption in French elderly community dwellers: data from the Three-City study. European Journal of Clinical Nutrition 59(7):817-25.

Beauchesne-Rondeau E, Gascon A, Bergeron J, Jacques H. 2003. Plasma lipids and lipoproteins in hypercholesterolemic men fed a lipid-lowering diet containing lean beef, lean fish, or poultry. Am J Clin Nutr 77(3):587-93.

Beil FU, Grundy SM. 1980. Studies on plasma lipoproteins during absorption of exogenous lecithin in man. J Lipid Res 21:525-536.

Biosca MDG, Fernandez-Cruz A, Nara Y, Yamori Y. 1990. Relationship between Urinary Amino-Acids and Diets in Spanish Cardiovascular-Diseases and Alimentary Comparison Study. Journal of Cardiovascular Pharmacology 16:32-34.

Boehm G, Borte M, Böhles HI, Müller H, Kohn G, Moro G. 1996. Docosahexaenoic and arachidonic acid content of serum and red blood cell membrane phospholipids of preterm infants fed breast milk, standard formula or formula supplemented with n-3 and n-6 long-chain polyunsaturated fatty acids. European Journal of Pediatrics 155(5):410-416.

Bordenave S, Fruitier I, Ballandier I, Sannier F, Gildberg A, Batista I, Piot J-M. 2002. HPLC preparation of fish waste hydrolysate fractions, effect on Guinea pig ileum and ACE activity. Prep. Biochem. Biotechnol 32:65-77.

Borzoei S, Neovius M, Barkeling B, Teixeira-Pinto A, Rössner S. 2006. A comparison of effects of fish and beef protein on satiety in normal weight men. Eur J Clin Nutr 60(7):897-902.

Bosetti C, Kolonel L, Negri E, Ron E, Franceschi S, Dal Maso L, Rosaria Galanti M, Mark SD, Preston-Martin S, McTiernan A, Land C,Jin F, Wingren G, Hallquist A, Glattre E, Lund E, Levi F, Linos D, La Vecchia C. 2001. A pooled analysis of case-control studies of thyroid cancer. VI. Fish and shellfish consumption. Cancer Causes and Control 12(4):375-82.

Bouckenooghe T, Remacle C, Reusens B. 2006. Is taurine a functional nutrient? Current Opinion in Clinical Nutrition and Metabolic Care 9(6):728-733.

Bourkott FO, Girard A, Prost J, Ait-Yahia D, Bouchenak M, Belleville J. 2004. Fish protein improves the total antioxidant status of steptozotocin-induced diabetes in spontaneously hypertensive rat. Medical Science Monitor 10:BR397-404.

Bouzan C, Cohen J, Connor W, Kris-Etherton P, Gray G, König A, Lawrence R, Savitz D, Teutsch S. 2005. A quantitative analysis of fish consumption and stroke risk. Am J Prev Med 29(4):347-52.

Breslow JL. 2006. n-3 fatty acids and cardiovascular disease. Am J Clin Nutr 83(6 Suppl):1477S-1482S.

Brøns C, Spohr C, Storgaard H, Dyerberg J, Vaag A. 2004. Effect of taurine treatment on insulin secretion and action, and on serum lipid levels in overweight men with a genetic predisposition for type II diabetes mellitus. European Journal of Clinical Nutrition 58(9):1239-1247.

Brown AA, Hu FB. 2001. Dietary modulation of endothelial function: implications for cardiovascular disease. American Journal of Clinical Nutrition 73(4):673-686.

Browne JC, Scott KM, Silvers KM. 2006. Fish consumption in pregnancy and omega-3 status after birth are not associated with postnatal depression. J Affect Disord 90(2-3):131-139.

Bruyere O, Reginster J-Y. 2007. Glucosamine and chondroitin sulphate as therapeutic agents for knee and hip osteoarthritis. Drugs Aging 24 (7):573-580.

Bulliyya G. 2000. Fish intake and blood lipids in fish eating vs non-fish eating communities of coastal south India. Clin Nutr 19(3):165-170.

Bulliyya G. 2002. Influence of fish consumption on the distribution of serum cholesterol in lipoprotein fractions: comparative study among fish-consuming and non-fish-consuming populations. Asia Pacific Journal of Clinical Nutrition 11(2):104-111.

Burr ML, Ashfield-Watt PAL, Dunstan FDJ, Fehily AM, Breay P, Ashton T, Zotos PC, Haboubi NAA, Elwood PC. 1989, Effects of changes in fat, fish, and fibre intakes on death and myocardial reinfarction: diet and reinfarction trial (DART). Lancet 2(8666):757-761.

Burr ML, Ashfield-Watt PAL, Dunstan FDJ, Fehily AM, Breay P, Ashton T, Zotos1 PC, Haboubi NAA, Elwood PC. 2003. Lack of benefit of dietary advice to men with angina: results of a controlled trial. Eur J Clin Nutr 57(2):193-200.

Burr ML, Fehily AM. 1991. Fatty fish and heart disease: a randomized controlled trial. World Rev Nutr Diet 66:306-12.

Burr ML. 1992. Fish food, fish oil and cardiovascular disease. Clin Exp Hypertens A 14(1-2):181-192.

Burr ML. 1993. Fish and ischaemic heart disease. World Rev Nutr Diet 72:49-60.

Byun HG, Kim YT, Park PJ, Lin X, Kim SK. 2005. Chitooligosaccharides as a novel beta-secretase inhibitor. Carbohyd. Polymers 61:198-202.

Cabanero AI, Carvahlo C, Madrid Y, Batoreu C, Cámara C. 2005. Quantification and speciation of mercury and selenium in fish samples of high consumption in Spain and Portugal. Biol Trace Elem Res 103:17-35

Calder PC. 2006. n-3 polyunsaturated fatty acids, inflammation, and inflammatory diseases. Am J Clin Nutr 83(6 Suppl):1505S-1519S.

Calvani M, Alessandri C, Miceli Sopo S, Panetta V, Pingitore J, Tripodi S, Zappalà D, Zicari AM. 2006. Consumption of fish, butter and margarine during pregnancy and development of allergic sensitizations in the offspring: role of maternal atopy. Pediatr Allergy Immunol 17(2):94-102.

Carnielli V, Verlato G, Perderzini F, Luijendijk I, Boerlage A, Pedrotti D, Sauer P. 1998. Intestional absorption of long-chain polyunsaturated fatty acids in preterm infants fed breast milk or formulae. American Journal of Clinical Nutrition 67:97-103.

Carpentier YA, Portois L, Malaisse WJ. 2006. n-3 fatty acids and the metabolic syndrome. Am J Clin Nutr 83(6 Suppl):1499S-1504S.

Chambers SJ, Wickham MS, Regoli M, Bertelli E, Gunning PA, Nicoletti C. 2004. Rapid *in vivo* transport of proteins from digested allergen across pre-sensitized gut. Biochem Biophys Res Commun 325:1258-1263.

Cheatham CL, Colombo J, Carlson SE. 2006. N-3 fatty acids and cognitive and visual acuity development: methodologic and conceptual considerations. Am J Clin Nutr 83(6 Suppl):1458S-1466S.

Chrysohoou C, Panagiotakos DB, Pitsavos C, Skoumas J, Krinos X, Chloptsios Y, Nikolaou V, Stefanadis C. 2007. Long-term fish consumption is associated with protection against arrhythmia in healthy persons in a Mediterranean region - the ATTICA study. Am J Clin Nutr 85(5):1385-1391.

Clandinin MT, Van Aerde JE, Merkel KL, Harris CL, Springer MA, Hansen JW, Diersen-Schade DA. 2005. Growth and development of preterm infants fed infant formulas containing docosahexaenoic acid and arachidonic acid. J Pediatr 146(4):461-468.

Clark LC,Combs Jr GF, Turnbull BW, Slate EH, Chalker DK, Chow J, Davis LS, Glover RA, Graham GF, Gross EG, Krongrad A, Lesher Jr LJ, Park HK, Sanders Jr BB, Smith CL, Taylor JR. 1996. Effects of selenium supplementation for cancer prevention in patients with carcinoma in the skin. J Am Med Assoc 276:1957-1985.

Clayton EH, Hanstock TL, Garg ML, Hazell PL. 2007. Long chain omega-3 polyunsaturated fatty acids in the treatment of psychiatric illnesses in children and adolescents. Acta Neuropsychiatrica 19(2):92-103.

Clegg DO, Reda DJ, Klein MA, O'Dell JR, Hooper MM, Bradley JD, Bingham CO, Weisman MH, Jackson CG, Lane NE, Cush JJ, Moreland LW, Schumacher Jr. HR, Oddis CV, Wolfe F, Molitor JA, Yocum DE, Schnitzer TJ, Furst DE, Sawitzke AD, Shi H, Brandt K, Moskowitz RW, Williams HJ. 2006. Glucosamine, Chondroitine sulfate, and the two in combination for painful knee osteoarthritis. New England Journal of Medicine 354(8):795-808.

Clement IP, Birringer M, Block E, Kotrebai M, Tyson JF, Uden PC, Lisk DJ. 2000. Chemical speciation influences comparative activity of selenium enriched garlic and yeast in mammary cancer prevention. J Agric Food Chem 48:2062-2070.

Cohen J, Bellinger D, Connor W, Kris-Etherton P, Lawrence R, Savitz D, Shaywitz B, Teutsch B, Gray G. 2005. A quantitative risk-benefit analysis of changes in population fish consumption. Am J Prev Med 28(4):325-334.

Collier PM, Ursell A, Zaremba K, Payne CM, Staughton RC, Sanders T. 1993. Effect of regular consumption of oily fish compared with white fish on chronic plaque psoriasis. Eur J Clin Nutr 47(4):251-254.

Collier PM, Payne CR. 1996. The dietary effect of oily fish consumption on psoriasis. Br J Dermatol 135(5):858.

Commission Regulation. 2006. Setting maximum levels for certain contaminants in foodstuffs. 2006. Official Journal of the European Union.

Combs GF. 2005. Current evidence and research needs to support a health claim for selenium and cancer prevention. J. Nutrition 335: 343-347.

Costa LG. 2007. Contaminants in Fish: Risk-Benefit Considerations. Arh Hig Rada Toksikol 58(3):367-374.

Cranney A, Horsley T, O'Donnell S, Weiler H, Puil L, Ooi D, Atkinson S, Ward L, Moher D. 2007. Effectiveness and safety of vitamin D in relation to bone health. Evidence Rep Technology Assessment (Full Rep) 158:1-235.

Crews HM. 1998. Speciation of trace elements in foods, with special reference to cadmium and selenium: is it necessary? Spectochmica Acta Part B 53:213-219.

Curtis GH, Gill DG. 1992. Macromolecular transport by rat gastric mucosa. American Journal of Physiology 262:G1033-1040.

Cutolo M, Otsa K, Uprus M, Paolino S, Seriolo B. 2007. Vitamin D in rheumatoid arthritis. Autoimmun Rev 7(1):59-64.

Dallongeville J, Yarnell J, Ducimetière P, Arveiler D, Ferrières J, Montaye M, Luc G, Evans A, Bingham A, Hass B, Ruidavets J-B, Amouyel P. 2003. Fish consumption is associated with lower heart rates. Circulation 108(7):820-825.

Daniels JL, Longnecker MP, Rowland AS, Golding J. 2004. Fish intake during pregnancy and early cognitive development of offspring. Epidemiology 15(4):394-402.

De Deckere EA. 1999. Possible beneficial effect of fish and fish n-3 polyunsaturated fatty acids in breast and colorectal cancer. Eur J Cancer Prev 8(3):213-221.

Demonty I, Deshaies Y, Lamarche B, Jacques H. 2003. Cod protein lowers the hepatic triglyceride secretion rate in rats. J Nutr 133(5):1398-1402.

Deslypere JP. 1990. Effect of fish consumption compared to intake of fish oil. Bibl Nutr Dieta 46(46):53-69.

Din J, Harding S, Valerio C, Sarma J, Lyall K, Riemersma R, Newby D, Flapan A. 2008. Dietary intervention with oil rich fish reduces platelet-monocyte aggregation in man. Atherosclerosis 197(1):290-296.

Distler J, Anguelouch A. 2006. Evidence-based practice: Review of the clinical evidence on the efficacy of glucosamine and chondroitin in the treatment of osteoarthritis. Journal of the American Academy of Nurse Practitioners 18:487-493.

Dong YL, Sheng G, Fu J, Wen K. 2005. Chemical characterization and anti-anaemia activity of fish protein hydrolysate from Saurida elongate. J Sci Food Agric 85:2033-2039.

Dragnes B, Larsen R, Holmen Ernstsen M, Maehre H, Elevevoll E. 2008. Impact of processing on taurine content in processed seafood and their corresponding unprocessed raw materials. International Journal of Food Science and Nutrition 12:1-10.

Duarte J, Vinderola G, Ritz B, Perdigón G, Matar C. 2006. Immunomodulating capacity of commercial fish protein hydrolysate for diet supplement. Immunobiol 211:341-350.

Dumont E, Vanhaecke F, Cornelis R. 2006. Selenium speciation from food source to metabolites: a critical review. Anal Bioanal Chem 385:1304-1323.

Dunstan DW, Mori TA, Puddey IB, Beilin LJ, Burke V, Morton AR, Stanton KG. 1997. The independent and combined effects of aerobic exercise and dietary fish intake on serum lipids and glycemic control in NIDDM. A randomized controlled study. Diabetes Care 20(6):913-921.

Dunstan DW, Morita A, Puddey LB, Beilin L J, Burke V, Morton AR, Stanton KG. 1999. A randomised, controlled study of the effects of aerobic exercise and dietary fish on coagulation and fibrinolytic factors in type 2 diabetics. Thromb Haemost 81(3):367-372.

EFSA Scientific Committee on Food. Opinion on Caffeine, Taurine and D-Glucurono - g -Lactone as constituents of so-called 'energy' drinks 1999 [cited; Available from: http://ec.europa.eu/food/fs/sc/scf/out22_en.html.

Eilander A, Hundscheid DC, Osendarp SJ, Transler C, Zock PL. 2007. Effects of n-3 long chain polyunsaturated fatty acid supplementation on visual and cognitive development throughout childhood: a review of human studies. Prostaglandins Leukot Essent Fatty Acids 76(4):189-203.

Elvevoll E. 2005. Fish waste and functional foods. In: Waldron K, Faulds C and Smith A (eds.) Total Food; Exploiting co-products; minimising Waste. Norwich: Institute for Food Research. p. 49-57.

Elvevoll EO, Österud B. 2003. Impact of processing on nutritional quality of marine food items. In: Elmadfa I AE, König JS (eds.) Modern aspects of marine food items-Present knowledge and future perspectives. Basel: Karger Press p. 337-341.

Elvevoll EO, Eilertsen KE, Brox J, Dragnes BT, Falkenberg P, Olsen JO, Kirkhus B, Lamglait A, Österud B. 2008. Seafood diets: Hypolipidemic and antiatherogenic effects of taurine and n-3 fatty acids. Atherosclerosis 200(2):396-402.

Engeset D, Alsaker E, Lund E, Welch A, Khaw K-T, Clavel-Chapelon F, Thiébaut A, Chajès V, Key TJ, Allen NE, Amiano P, Dorronsoro P, Tjønneland A, Stripp C, Peeters PHM, van Gils CH, Chirlaque M-D, Nagel G, Linseisen J, Ocké MC, Bas Bueno-de-Mesquita H, Sacerdote C, Tumino R, Ardanaz E, Sánchez M-J, Panico S, Palli D, Trichopoulou A, Kalapothaki V, Benetou V, Ramón Quirós J, Agudo A, Overvad K, Bjerregaard L, Wirfält E, Schulz M, Boeing H, Slimani N, Riboli E. 2006. Fish consumption and breast cancer risk. The European Prospective Investigation into Cancer and Nutrition (EPIC). Int J Cancer 119(1):175-82.

Engstrom K, Wallin R, Saldeen T. 2003. Effects of Scandinavian caviar paste enriched with a stable fish oil on plasma phospholipid fatty acids and lipid peroxidation. Eur J Clin Nutr 57(9):1052-1059.

Erkkila AT, Lichtenstein AH, Mozaffarian D, Herrington DM. 2004. Fish intake is associated with a reduced progression of coronary artery atherosclerosis in postmenopausal women with coronary artery disease. Am J Clin Nutr 80(3):626-632.

Fahmi A, Morimura S, Guo HC, Shigematsu T, Kida K, Uemura Y. 2004. Production of angiotensin I converting enzyme inhibitory peptides from sea bream scales. Process Biochem 39:1195-1200.

Fang PC, Kuo HK, Huang CB, Ko TY, Chen CC, Chung MY. 2005. The effect of supplementation of docosahexaenoic acid and arachidonic acid on visual acuity and neurodevelopment in larger preterm infants. Chang Gung Med J 28(10):708-715.

FAO. 1970. Amino acid content of foods and biological data of proteins (2nd ed.). FAO Nutritional Study no. 24. Rome: Food Science Service.

FAO/WHO/UNU Joint Expert Consultation. 1985. Energy and protein requirements. WHO Geneva.

FAOSTAT, 2001 FAOSTAT, FAO statistical databases, fisheries data. 2001. Food and Agriculture Organization of the United Nations, Rome, Italy. Available from http://www.fao.org.

Fave G, Coste TC, Armand M. 2004. Physiochemical properties of lipids: New strategies to manage fatty acid bioavailability. Cellular and Molecular biology 50(7):815-831.

Fernandez E, Chatenoud L, La Vecchia C, Negri E, Franceschi S. 1999. Fish consumption and cancer risk. Am J Clin Nutr 70(1): 85-90.

Fewtrell MS, Abbott RA, Kennedy K, Singhal A, Morley R, Caine E, Jamieson C, Cockburn F, Lucas A. 2004. Randomized, double-blind trial of long-chain polyunsaturated fatty acid supplementation with fish oil and borage oil in preterm infants. J Pediatr 144(4):471-479.

Finnley JW, Davis CD, Feng Y. 2000. Selenium from high selenium broccoli protects rats from colon cancer. Journal of nutrition 130:2384-2389.

Fox TE, Van den Heuvel EGHM, Atherton CA, Dainty JR, Lewis DJ, Langford NJ, Crews HM, Luten JB, Lorentzen M, Sieling FW, van Aken-Schneyder P, Hoek M, Kotterman MJM, van Dael P, Fairweather-Tait SJ. 2004. Bioavailability of selenium from fish, yeast and selenate: a comparative study in humans using stable isotopes. European Journal of Clinical Nutrition 58:343-349.

Friedland RP. 2003. Fish consumption and the risk of Alzheimer disease: is it time to make dietary recommendations? Arch Neurol 60(7):923-924.

Friedman M. 1996. Nutritional value of proteins from different food sources. A review. Journal of Agricultural and Food Chemistry 44(1):6-29.

Fujimoto T, Tsuchiya Y, Terao M, Nakamura K, Yamamoto M. 2006. Antibacterial effects of chitosan solutions against Legionella pneumophila, Escherichia coli, and Staphylococcus aureus. Internat J Food Microbiol 112:96-101.

Gama CI, Hsieh-Wilson LC. 2005. Chemical approaches to deciphering the glycosaminoglycan code. Current opinion in Chemical Biology 9:609-619.

Garland CF, Garland FC, Gorham ED, Lipkin M, Newmark H, Mohr SB, Hollick MF. 2006. The role of vitamin D in cancer prevention. Am J Public Health 96(2):252-61.

Gildberg A, Bøgwald J, Johansen A, Stenberg E. 1996. Isolation of acid peptide fractions from fish protein hydrolysate with strong stimulatory effect on Atlantic salmon (Salmo salar) head kidney leucocytes. Comp Biochem Physiol 114B:97-101.

Gloth, FM, Alam W, Hollis B. 1999. Vitamin D vs broad spectrum phototherapy in the treatment of seasonal affective disorder. J Nutr Health Aging 3(1):5-7.

Gochfeld M, Burger J. 2005. Good fish/bad fish: a composite benefit-risk by dose curve. Neurotoxicology 26(4):511-520.

Goggs R, Vaughan-Thomas A, Clegg PD, Carter SD, Innes JF, Mobasheri A, Shakibaei M, Schwab W, Bondy CA. 2005. Nutriceutical therapies for degenerative joint diseases: A critical review. Critical Reviews in Food Science and Nutrition 45:145-164.

Gorham ED, Garland CF, Garland FC, Grant WB, Mohr SB, Lipkin M, Newmark HL, Giovannucci E, Wei M, Holick MF. 2005. Vitamin D and prevention of colorectal cancer. J Steroid Biochem Mol Biol 97(1-2):179-194.

Guo Z, Vikbjerg F, Xu X. 2005. Enzymatic modification of phospholipids for functional applications and human nutrition. Biotechnology Advances 23(3):203-259.

Hagmar L, Persson-Moschos M, Akkesson B, Schuetz A. 1998. Plasma levels of selenium, selenoprotein P and glutathione peroxidase and their correlations to fish intake and serum levels to thyrotropin and thyroid hormones: a study on Latvian fish consumers. European Journal of Clinical Nutrition 52:796-800.

Halton TL, Hu FB. 2004. The effects of high protein diets on thermogenesis, stiety and weight loss: a critical review. Journal of the American College of Nutrition 23:373-385

Hamer M, Steptoe A. 2006. Influence of specific nutrients on progression of atherosclerosis, vascular function, haemostasis and inflammation in coronary heart disease patients: a systematic review. Br J Nutr 95(5):849-859.

Han LK, Sumiyoshi M, Takeda T, Chihara H, Nishikiori T, Tsujita T, Kimura Y, Okuda H. 2000. Inhibitory effects of chondroitin sulfate prepared from salmon nasal cartilage on fat storage in mice fed a high-fat diet. International Journal of Obesity 24:1131-1138.

Han YP, Zhao LH, Yu ZJ, Feng J, Yu QQ. 2005. The role of mannose receptor in oligochitosan-mediated stimulation of macrophage function. Internat. Immunopharmacol 5:1533-1542.

Harris WS. 2006. The omega-6/omega-3 ratio and cardiovascular disease risk: uses and abuses. Curr Atheroscler Rep 8(6):453-459.

He K, Rimm EB, Merchant A, Rosner BA, Stampfer MJ, Willett WC, Ascherio A. 2002. Fish consumption and risk of stroke in men. Jama 288(24):3130-3136.

He K, Song Y, Daviglus ML, Liu K, Van Horn, Dyer AR, Greenland P. 2004a. Accumulated evidence on fish consumption and coronary heart disease mortality: a meta-analysis of cohort studies. Circulation 109(22):2705-2711.

He K, Song Y, Daviglus ML, Liu K, Van Horn L, Dyer AR, Goldbourt U, Greenland P. 2004b. Fish consumption and incidence of stroke: a meta-analysis of cohort studies. Stroke 35(7):1538-1542.

He H-L, Xiu-Lan C, Cai-Yun S, Yu-Zhong Z, Bai-Cheng Z. 2006. Analysis of novel angiotensin-I-converting enzyme inhibitory peptides from protease-hydrolyzed marine shrimp Acetes chinensis. J Pept Sci 12:726-733.

He K, Daviglus ML. 2005. A few more thoughts about fish and fish oil. Journal of the American Dietetic Association 105(3):350-351.

Hibbeln JR, Nieminen LRG, Blasbalg TL, Riggs JA, Lands WEM. 2006. Healthy intakes of n-3 and n-6 fatty acids: estimations considering worldwide diversity. Am J Clin Nutr 83(6 Suppl):1483S-1493S.

Hibbeln JR, Davis JM, Steer C, Emmett P, Rogers I, Williams C, Golding J. 2007. Maternal seafood consumption in pregnancy and neurodevelopmental outcomes in childhood (ALSPAC study): an observational cohort study. Lancet 369(9561):578-585.

Higdon JV, Du SH, Lee YS, Wu T, Wander RC. 2001. Supplementation of postmenopausal women with fish oil does not increase overall oxidation of LDL ex vivo compared to dietary oils rich in oleate and linoleate. J Lipid Res 42:407-418.

Higdon JV, Liu J, Du SH, Morrow JD, Ames BN, Wander RC. 2000. Supplementation of postmenopausal women with fish oil rich in eicosapentaenoic acid and docosahexaenoic acid is not associated with greater *in vivo* lipid peroxidation compared with oils rich in oleate and linoleate as assessed by plasma malondialdehyde and F(2)-isoprostanes. Am J Clin Nutr 72:714-722.

Higgins S, Carroll YL, McCarthy SN, Corridan BM, Roche HM, Wallace JMW, O'Brian NM, Morrissey PA. 2001. Susceptibility of LDL to oxidative modification in healthy volunteers supplemented with low doses of n-3 polyunsaturated fatty acids. Br J Nutr 85:23-31.

Holdway DA, Beamish FWH. 1984. Specific growth rate and proximate body composition of Atlantic cod (Gadus morhua). J Exp Mar Biol Ecol 81:147-170.

Holick MF. 2007. Vitamin D Deficiency. N Engl J Med 357:266-281.

Holick MF, Chen TC. 2008. Vitamin D deficiency: a worldwide problem with health concequences. Am J Clin Nutr 87:1080S-1086S.

Hooper L, Thompson RL, Harrison RA, Summerbell CD, Ness AR, Moore HJ, Worthington HV, Durrington PN, Higgins JPT, Capps NE, Riemersma RA, Ebrahim SBJ, Smith GD. 2006. Risks and benefits of omega 3 fats for mortality, cardiovascular disease, and cancer: systematic review. BMJ 332(7544):752-760.

Howling GI. 2002. The effect of chitin and chitosan on fibroblast-populated collagen lattice contraction. Biotechnol Appl Biochem 36:247-253.

Hu FB, Cho E, Rexrode KM, Albert CM, Manson JE. 2002. Fish and long-chain omega-3 fatty acid intake and risk of coronary heart disease and total mortality in diabetic women. Circulation 107(14):1852-1857.

Huang TL, Zandi PP, Tucker KL, Fitzpatrick AL, Kuller LH, Fried LP, Burke GL, Carlson MC. 2005. Benefits of fatty fish on dementia risk are stronger for those without APOE epsilon4. Neurology 65(9):1409-1414.

Hunter DJ, Kazda I, Chockalingam A, Fodor JG. 1988. Fish consumption and cardiovascular mortality in Canada: an inter-regional comparison. American Journal of Preventive Medicine 4(1):5-10.

Hurley C, Galibois I, Jacques H. 1995. Fasting and postprandial lipid and glucose metabolisms are modulated by dietary proteins and carbohydrates: Role of plasma insulin concentrations. Nutr Biochem 6:540-546.

Ichimura T, Hu J, Aita DQ, Maruyama S. 2003. Angiotensin I-converting enzyme inhibitory activity and insulin stimulative activity of fermented fish sauce. J Biosci Bioeng 96:496-499.

Interim Report 5 Study of dioxin like PCBs in fatty fish from Sweden 2000-2002. 2002. Uppsala: Swedish National Food Administration.

Iso H, Rexrode KM, Stampfer MJ, Manson JE,. Colditz GA, Speizer FE, Hennekens CH, Willett WC. 2001. Intake of fish and omega-3 fatty acids and risk of stroke in women. JAMA 285(3):304-312.

Iso H, Kobayashi M, Ishihara J, Sasaki S, Okada K, Kita Y, Kokubo Y, Tsugane S. 2006. Intake of fish and n3 fatty acids and risk of coronary heart disease among Japanese: the Japan Public Health Center-Based (JPHC) Study Cohort I. Circulation 113(2):195-202.

Jensen CL. 2006. Effects of n-3 fatty acids during pregnancy and lactation. Am J Clin Nutr 83(6 Suppl):1452S-1457S.

Jeon YJ, Kim SK. 2002. Antitumor activity of chitosan oligosaccharides produced in ultrafiltration membrane reactor system. J Microbiol Biotechnol 12:503-507.

Johnson EJ, Schaefer EJ. 2006. Potential role of dietary n-3 fatty acids in the prevention of dementia and macular degeneration. Am J Clin Nutr 83(6 Suppl):1494S-1498S.

Jung WK. 2006. Angiotensin I-converting enzyme inhibitory peptide from yellowfin sole (Limanda aspera) frame protein and its antihypertensive effect in spontaneously hypertensive rats. Food Chem 94:26-32.

Kalmijn S, van Boxtel MPJ, Ocké M, Verschuren WMM, Kromhout D, Launer LJ. 2004. Dietary intake of fatty acids and fish in relation to cognitive performance at middle age. Neurology 62(2):275-280.

Kalmijn S. 2000. Fatty acid intake and the risk of dementia and cognitive decline: a review of clinical and epidemiological studies. J Nutr Health Aging 4(4):202-207.

Kalueff AV, Eremin KO, Tuohimaa P. 2004. Mechanisms of neuroprotective action of vitamin D3. Biochemistry (Mosc.) 69(7):738-41

Kawasaki T, Jun CJ, Fukushima Y, Kegai K, Seki E, Osajima K, Itoh K, Matsui T, Matsumoto K. 2002. Antihypertensive effect and safety evaluation of vegetable drink with peptides derived from sardine protein hydrolysates on mild hypertensive, high-normal and normal blood pressure subjects. Fukuoka Igaku Zasshi 93(10):208-218.

Kibayashi E, Yokogoshi H, Mizue H, Miura K, Yoshita K, Nakagawa H, Naruse Y, Sokejima S, Kagamimori S. 2000. Daily dietary taurine intake in Japan. Adv Exp Med Biol 483:137-142.

Kim ES, Kim JS, Yim MH, Jeong Y, Ko YS, Watanabe T, Nakatsuka H, Nakatsuka S, Matsuda-Inoguchi N, Shimbo S, Ikeda M. 2003. Dietary taurine intake and serum thurine levels of women on Jeju Island. Adv Exp Med Biol 277-283.

Kim SK, Kim YT, Byun HG, Nam KS, Joo DS, Shahidi F. 2001. Isolation and characterization of antioxidative peptides from gelatine hydrolysate of Alaska Pollack skin. J Agric Food Chem 49:1984-1989.

Kim SK, Mendis E. 2006. Bioactive compounds for marine processing byproducts – A review. Food Res Internat 39:383-393.

Kimura Y, Kono S, Toyomura K, Nagano J, Mizoue T, Moore MA, Mibu R, Tanaka M, Kakeji Y, Maehara Y, Okamura T, Ikejiri K, Futami K, Yasunami Y, Maekawa T, Takenaka K, Ichimiya H, Imaizumi N. 2007. Meat, fish and fat intake in relation to subsite-specific risk of colorectal cancer: The Fukuoka Colorectal Cancer Study. Cancer Sci 98(4):590-597.

Klein EA, Thompson IM, Lippman SM, Goodman PJ, Albanes D, Taylor PR, Coltman C. 2000. SELECT: the Selenium and Vitamin E Cancer Prevention Trial: rationale and design, Prostate Cancer and Prostatic Diseases 3:45-151.

Knapp HR. 1997. Dietary fatty acids in human thrombosis and hemostasis. Am J Clin Nutr 65(5 Suppl):1687S-1698S.

Knekt P. 2002. Selenium status and prevention of chronic diseases, In Cadenas E, Packer L Ediotors, Handbook of antioxidants, New York and Basel, Marcel Dekker Inc, 665-687.

Kobayashi M, Tsubono Y, Otani T, Hanaoka T, Sobue T, Tsugane S. 2004. Fish, long-chain n-3 polyunsaturated fatty acids, and risk of colorectal cancer in middle-aged Japanese: the JPHC study. Nutrition and Cancer 49(1):32-40.

Kolanowski W, Swiderski F, Berger S. 1999. Possibilities of fish oil application for food products enrichment with omega-3 PUFA. Int J Food Sci Nutr 50(1):39-49.

König A, Bouzan C, Cohen J, Connor W, Kris-Etherton P, Gray G, Lawrence R, Savitz D, Teutsch S. 2005. A quantitative analysis of fish consumption and coronary heart disease mortality. Am J Prev Med 29(4):335-346.

Kris-Etherton P, Harris WS, Appel LJ. 2002. AHA Scientific Statement: Fish consumption, fish oil, omega-3 fatty acids, and cardiovascular disease. Circulation Nov 19:106(21):2747-2757.

Kris-Etherton PM, Shaffer Taylor D, Yu-Poth S, Huth P, Moriarty K, Fishell V, Hargrove RL, Zhao G, Etherton TD. 2000. Polyunsaturated fatty acids in the food chain in the United States. Am J Clin Nutr 71:179S-188S.

Kristensen SD, Iversen AM, Schmidt EB. 2001. n-3 polyunsaturated fatty acids and coronary thrombosis. Lipids 36 Suppl:S79-82.

Kromhout, D, Bosschieter EB, de Lezenne Coulander C. 1985. The inverese relation between fish consumption and 20-year mortality from coronary heart disease. The New England Journal of Medicine 312(19):1205-1209.

Kull I, Bergstrom A, Lilja G, Pershagen G, Wickman M. 2006. Fish consumption during the first year of life and development of allergic diseases during childhood. Allergy 61(8):1009-1015.

Lacaille B, Julien P, Deshaies Y, Lavigne C, Brun L-D, Jacques H. 2000. Responses of plasma lipoproteins and sex hormones to the consumption of lean fish incorporated in a prudent-type diet in normolipidemic men. J Am Coll Nutr 19(6):745-753.

Lakhan SE, Vieira KF. 2008. Nutritional therapies for mental disorders. Nutrition Journal 7:2.

Lara JJ, Economou M, Wallace A, Rumley A, Lowe G, Slater C, Caslake M, Sattar N, Lean M. 2006. Benefits of salmon eating on traditional and novel vascular risk factors in young, non-obese healthy subjects. Atherosclerosis 25:25.

Lavigne C, Marette A, Jacques H. 2000. Cod and soy proteins compared with casein improve glucose tolerance and insulin sensitivity in rats. American Journal of Physiology 278:E491-500.

Lavigne C, Tremblay F, Asselin G, Jacques H, Marette A. 2001. Prevention of skeletal muscle insulin resistance by dietary cod protein in high fat-fed rats. American Journal of Physiology 281:E62-71.

Leeb B, Schweitzer H, Montag K, Smolen J. 2000. A metaanalysis of chondroitin sulfate in the treatment of osteoarthritis. Journal of Rheumatology 27:205-211.

Lemaitre-Delaunay D, Pachiaudi C, Laville M, Pousin J, Armstrong M, Lagarde M. 1999. Blood compartmental metabolism of docosahexaenoic acid (DHA) in humans after ingestion of a single dose of [13C]DHA in phosphatidylcholine, Journal of Lipid Research, 40(10):1867-1874.

Lewin GA, Schachter HM, Yuen D, Merchant P, Mamaladze V, Tsertsvadze A. 2005. Effects of omega-3 fatty acids on child and maternal health. Evid Rep Technol Assess (Summ). 118:1-11. Review

Li Z, Lamon-Fava S, Otvos J, Lichtenstein AH, Velez-Carrasco W, McNamara JR, Ordovas JM, Schaefer EJ. 2004 Fish consumption shifts lipoprotein subfractions to a less atherogenic pattern in humans. J Nutr 134(7):1724-1728.

Lindqvist H, Langkilde A-M, Undeland I, Rådendal T, Sandberg A-S. 2007. Herring (*Clupea harengus*) supplemented diet influence risk factors for CVD in verweight subjects. Eur J Clin Nutr 61(9):1106-1113.

Lindqvist H, Langkilde A-M, Undeland I, Sandberg A-S. 2008. Herring (*Clupea harengus*) intake influence lipoproteins but not inflammation and oxidation markers in overweight men. Br J Nutr 18:1-8.

Liu H, Du Y, Wang X, Sun L. 2004a. Chitosan kills bacteria through cell membrane damage. Internat. J Food Microbiol 95:147-155.

Liu F, Iqbal K, Grundke-Iqbal I, Hart GW, Gong C-X. 2004b. O-Glc Nacylation regulates phosphorylation of tau: A mechanism involved in Alzheimer's disease. Proc Natl Acad Sci USA 101:10804-10809.

London B, Albert C, Anderson ME, Giles WR, Van Wagoner DR, Balk E, Billman GE, Chung M, Lands W, Leaf A, McAnulty J, Martens JR, Costello RB, Lathrop DA. 2007. Omega-3 Fatty acids and cardiac arrhythmias: prior studies and recommendations for future research: a report from the National Heart, Lung, and Blood Institute and Office Of Dietary Supplements Omega-3 Fatty Acids and Their Role In Cardiac Arrhythmogenesis Workshop. Circulation 116(10):e320-335.

Lovegrove JA, Brooks CN, Murphy MC, Gould BJ, Williams CM. 1997. Use of manufactured foods enriched with fish oils as a means of increasing long-chain n-3 polyunsaturated fatty acid intake. Br J Nutr 78:223-236.

Lüchtenborg M, Weijenberg MP, de Goeij AFPM, Wark PA, Brink M, Roemen GMJM, Lentjes MHFM, de Bruïne AM, Goldbohm RA, van 't Veer P, van den Brandt PA. 2005. Meat and fish consumption, APC gene mutations and hMLH1 expression in colon and rectal cancer: a prospective cohort study (The Netherlands). Cancer Causes and Control 16(9):1041-1054.

Lund E. 2002. Fish and cancer. IARC Sci Publ 156. p. 187-189.

MacLean CH, Newberry SJ, Mojica WA, Khanna P, Issa AM, Suttorp MJ, Lim Y-W, Traina SB, Hilton L, Garland R, Morton SC. 2006 Effects of omega-3 fatty acids on cancer risk: a systematic review. JAMA 295(4):403-415.

Mahadik SP, Pillai A, Joshi S, Foster A. 2006. Prevention of oxidative stress-mediated neuropathology and improved clinical outcome by adjunctive use of a combination of antioxidants and omega-3 fatty acids in schizophrenia. Int Rev Psychiatry 18(2):119-131.

Makrides M, Neumann MA, Byard RW, Simmer K, Gibson RA. 1994. Fatty acid composition of brain, retina and erythrocytes in breast and formula fed infants. American Journal of Clinical Nutrition 60:189-194.

Mantzioris E, Cleland LG, Gibson RA, Neumann MA, Demasi M, James MJ. 2000. Biochemical effects of a diet containing foods enriched with n-3 fatty acids. Am J Clin Nutr 72:42-48.

Marckmann P, Gronbaek M. 1999. Fish consumption and coronary heart disease mortality. A systematic review of prospective cohort studies. Eur J Clin Nutr 53(8):585-590.

Martins D, Wolf M, Pan D, Zadshir A, Tareen N, Thandhani R, Felsenfeld A, Levine B, Mehrotra R, Norris K. 2007. Prevalence of cardiovascular risk factors and the serum levels of 25-hydroxyvitamin D in the United States. Data from the Third National Health and Nutrition Examination Survey. Arch Intern Med 167:1159-1165.

Mastellone II, Polichetti E, Gres S, de la Maisonneuve C, Domingo N, Marin VV, Lorec A, Farnarier C, Portugal H, Kaplanski G, Chanussot F. 2000. Dietary soybean phosphatidylcholines lower lipidemia: mechanisms at the levels of intestine, endothelial cell, and hepato-biliary axis. J Nutr Biochem 11:461-466.

Matsumura N, Fujii M, Takeda Y, Sugita K, Shimizu T. 1993. Angiotensin I-converting enzyme inhibitory peptides derived from bonito bowels autolysate. Biosci Biotech Biochem 57:695-697.

McAlindon TE, LaValley MP, Gulin JP, Felson DT. 2000. Glucosamine and chondroitin for treatment of osteoarthritis – A systematic quality assessment and meta-analysis, Journal of the American Medical Association 283(11):1469-1475.

McGrath JJ, Feron FP, Burne TH, Mackay-Sim A, Eyles DW. 2004. Vitamin D3-implications for brain development. J Steriod Biochem Mol Biol 89-90 (1-5):557-560.

Meisel H, Schlimme E. 1996. Bioactive peptides derived from milk proteins: ingredients for functional foods? Kieler Milchwirtshaft Forschungsber 48:343-357.

Merrill AH, Symilen H, Duan R-D. 2003. Nutritional implications of sphingolipids: Occurrence and roles in cell regulation. In: Szuhaj BF, van Nieuwenhuyzen W (eds.) Nutrition and biochemistry of phospholipids. Champaign Illinois: AOCS Press. p. 61-69.

Metcalf RG, James MJ, Mantzioris E, Cleland LG. 2003. A practical approach to increasing intakes of n-3 polyunsaturated fatty acids: use of novel foods enriched with n-3 fats. Eur J Clin Nutr 57(12):1605-1612.

Michos ED, Melamed ML. 2008. Vitamin D and cardiovascular disease risk. Curr Opin Clin Nutr Metab Care 11(1):7-12.

Militante JD, Lombardini JB. 2004. Dietary taurine supplementation: hypolipidernic and antiatherogenic effects. Nutrition Research 24(10):787-801.

Miller DL. 2003. The benefits of lecithin on cardiovascular disease. In Szuhaj BF, van Nieuwenhuyzen W (eds.) Nutrition and biochemistry of phospholipids. Champaign Illinois: AOCS Press p. 117-129.

Miyake Y, Sasaki S, Yokoyama T, Tanaka K, Ohy Y, Fukushima W; Saito K, Ohfuji S, Kiyohara C, Hirota Y. 2006. Risk of postpartum depression in relation to dietary fish and fat intake in Japan: the Osaka Maternal and Child Health Study. Psychol Med 36(12):1727-1735.

Mizushima S, Moriguchi EH, Ishikawa P, Hekman P, Nara Y, Mimura G, Moriguchi Y, Yamori Y. 1997. Fish intake and cardiovascular risk among middle-aged Japanese in Japan and Brazil. Journal of Cardiovascular Risk 4(3):191-199.

Moreira AC, Gaspar A, Serra MA, Simões J, Lopes da Cruz J, Amaral TF. 2007. Effect of a sardine supplement on C-reactive protein in patients receiving hemodialysis. J Ren Nutr 17(3):205-213.

Moreno P, Quijano MA, Gutiérrez AM, Pérez-Conde MC, Cámara C. 2004. Study of selenium species distribution in biological tissues by size exclusion and ion exchange chromatography inductively coupled plasma-mass spectrometry, Analytica Chimica Acta 524:315-327.

Mori TA, Bao DQ, Burke V, Puddey IB, Watts GF, Beilin LJ. 1999a. Dietary fish as a major component of a weight-loss diet: effect on serum lipids, glucose, and insulin metabolism in overweight hypertensive subjects. Am J Clin Nutr 70(5):817-825.

Mori TA, Dunstan DW, Burke V, Croft KD, Rivera JH, Beilin LJ, Puddey IB. 1999b. Effect of dietary fish and exercise training on urinary F2-isoprostane excretion in non-insulin-dependent diabetic patients. Metabolism 48(11):1402-1408.

Mori TA, Burke V, Puddey IB, Shaw J, Beilin LJ. 2004. Effect of fish diets and weight loss on serum leptin concentration in overweight, treated-hypertensive subjects. J Hypertens 22(10):1983-1990.

Mori TA. 2006. Omega-3 fatty acids and hypertension in humans. Clin Exp Pharmacol Physiol. 33(9):842-846.

Morris MC, Evans DA, Bienias JL, Tangney CC, Bennett DA, Wilson RS, Aggarwal N, Schneider J. 2003. Consumption of fish and n-3 fatty acids and risk of incident Alzheimer disease. Arch Neurol 60(7):940-946.

Morris MC, Evans DA, Tangney CC, Bienias JL, Wilson RS. 2005. Fish consumption and cognitive decline with age in a large community study. Arch Neurol 62(12):1849-1853.

Mozaffarian D, Lemaitre RN, Kuller LH, Burke GL, Tracy RP, Siscovick DS. 2003. Cardiac benefits of fish consumption may depend on the type of fish meal consumed: the Cardiovascular Health Study Circulation, Circulation 107(10):1372-1377.

Mozaffarian D, Psaty BM, Rimm EB, Lemaitre RN, Burke GL, Lyles MF, Lefkowitz D, Siscovick DS. 2004. Fish intake and risk of incident atrial fibrillation. Circulation 110(4):368-373.

Mozaffarian D, Longstreth WT, Lemaitre RN, Manolio TA, Kuller LH, Burke GL, Siscovick DS. 2005a. Fish consumption and stroke risk in elderly individuals: the cardiovascular health study. Arch Intern Med 165(2):200-206.

Mozaffarian D, Prineas RJ, Stein PK, Siscovick DS. 2006. Dietary fish and n-3 fatty acid intake and cardiac electrocardiographic parameters in humans. J Am Coll Cardiol 48(3):478-484.

Mozaffarian D, Rimm EB. 2006. Fish intake, contaminants, and human health: evaluating the risks and the benefits. Jama 296(15):1885-1899.

Muller AD, van Houwelingen AC, van Dam-Mieras MC, Bas BM, Hornsyar G. 1989. Effect of a moderate fish intake on haemostatic parameters in healthy males. Thromb Haemost 61(3):468-473.

Muzzarelli RA. 1999. Clinical and biochemical evaluation of chitosan for hypercholesterolemia and overweight control. EXS. 293-304.

Myint PK, Welcha AA, Binghama SA, Lubena RN, Warehama NJ, Daya NE, Khawa K-T. 2006. Habitual fish consumption and risk of incident stroke: the European Prospective Investigation into Cancer (EPIC)-Norfolk prospective population study. Public Health Nutr 9(7):882-888.

Nakamura Y, Ueshima H, Okamura T, Kadowaki T, Hayakawa T, Kita Y, Tamaki S, Okayama A. 2005. Association between fish consumption and all-cause and cause-specific mortality in Japan: NIPPON DATA 80, 1980-99. The American Journal of Medicine 118(3):239-245.

Nandini CD, Itoh N, Sugahara K. 2005. Novel 70-kDa chondroitin/dermatan sulphate hybrid chains with a unique heterogenous sulfation pattern from shark skin, which exhibit neuritogenic activity and binding activities for growth factors and neurotrophic factors. J Biol Chem 280(6):4058-4069.

Nasopoulou C, Nomikos T, Demopoulos CA, Zabetakis I. 2007. Comparison of antiatherogenic properties of lipids obtained from wild and cultured sea bass (*Dicentrarchus labrax*) and gilthead sea bream (*Sparus aurata*). Food Chemistry 100(2):560-567.

National Academy of Sciences. 2000. Dietary Reference Intakes for Vitamin C, Vitamin E, Selenium and Carotenoids, National Academy Press, Washington, D.C.

Nelson GJ, Morris VC, Schmidt PC, Levander O. 1993. The urinary excretion of thiobarbituric acid reactive substances and malondialdehyde by normal adult males after consuming a diet containing salmon. Lipids 28(8):757-761.

Ness AR, Gallacher JE, Bennett PD, Gunnell DJ, Rogers PJ, Kessler D, Burr ML. 2003. Advice to eat fish and mood: a randomised controlled trial in men with angina. Nutritional Neuroscience 6(1):63-65.

Nestel P. 2001. Effects of fish oils and fish on cardiovascular disease. Curr Atheroscler Rep, 3(1):68-73.

Niittynen L, Nurminen ML, Korpela R, Vapaatalo H. 1999. Role of arginine, taurine and homocysteine in cardiovascular diseases. Annals of Medicine 31(5):318-326.

Nkondjock A, Receveur O. 2003. Fish-seafood consumption, obesity, and risk of type 2 diabetes: an ecological study. Diabetes Metab 29(6):635-642.

Norat T, Bingham S, Ferrari P, Slimani N, Jenab M, Mazuir M, Overvad K, Olsen A, Tjønneland A, Clavel F, Boutron-Ruault M-C, Kesse E, Boeing H, Bergmann MM, Nieters A, Linseisen J, Trichopoulou A, Trichopoulos D, Tountas Y, Berrino F, Palli D, Panico S, Tumino R, Vineis P, Bas Bueno-de-Mesquita H, Peeters PHM, Engeset D, Lund E, Skeie G, Ardanaz E, González C, Navarro C, Ramón Quirós J, Sanchez M-J, Berglund G, Mattisson I, Hallmans G, Palmqvist R, Day NE, Khaw K-T, Key TJ, San Joaquin M, Hémon B, Saracci R, Kaaks R, Riboli E. 2005. Meat, fish, and colorectal cancer risk: the European Prospective Investigation into cancer and nutrition. J Natl Cancer Inst 97(12):906-916.

Okamoto Y, Yano R, Miyatake K, Tomohiro I, Shigemasa Y, Minami S. 2003. Effects of chitin/chitosan and their oligomers/monomers on migration of macrophages. Macromol Biosci 3:587-590.

Olsen SF, Østerdal ML, Dalby-Salvig J, Kesmodel U, Brink-Henriksen T, Hedegaard M, Secher NJ. 2006. Duration of pregnancy in relation to seafood intake during early and mid pregnancy: prospective cohort. Eur J Epidemiol 21(10):749-758.

Önning G, Bergdahl IA. 1999. Fractionation of soluble selenium compounds from fish using siz-exclusion chromatography with on-line detection by inductively coupled plasma mass spectrometry. Analyst 124:1435-1438.

Oudit GY, Trivieri MG, Khaper N, Husain T, Wilson GJ, Liu P, Sole MJ, Backx PH. 2004. Taurine supplementation reduces oxidative stress and improves cardiovascular function in an iron-overload murine model. Circulation 109(15):1877-1885.

Ouellet V, Marois J, Weisnagel SJ, Jacques H. 2007. Dietary cod protein improves insulin sensitivity in insulin-resistant men and women: a randomized controlled trial. Diabetes Care 30(11):2816-2821.

Panagiotakos DB, Pitsavos C, Zampelas A, Chrysohoou C, Griffin B, Stefanadis C, Toutouzas P. 2005a. Fish consumption and the risk of developing acute coronary syndromes: the CARDIO2000 study. Int J Cardiol 102(3):403-409.

Panagiotakos DB, Pitsavos C, Zampleas A, Chrysohoou C, Stefanadis C. 2005b. The relationship between fish consumption and the risk of developing acute coronary syndromes among smokers: the CARDIO2000 case-control study. Nutr Metab Cardiovasc Dis 15(6):402-409.

Papakonstantinou E, Panagoitakos DB, Pitsavos C, Chysohoou C, Zampelas A, Skoums Y, Stefanadis C. 2005. Food group consumption and glycemic control in people with and without type 2 diabetes. Diabetes Care 28:2539-2540.

Pattison DJ, Harrison RA, Symmons DP. 2004. The role of diet in susceptibility to rheumatoid arthritis: a systematic review. J Rheumatol 31(7):1310-1319.

Pedersen M, Stripp C, Klarlund M, Olsen SF, Tjønneland AM, Frisch M. 2005. Diet and risk of rheumatoid arthritis in a prospective cohort. J Rheumatol 32(7):1249-1252.

Pepeu G. 2003. Is there evidence that phsopholipid administration is beneficial to your brain? In: Szuhaj BF, van Nieuwenhuyzen W (eds.) Nutrition and biochemistry of phospholipids. Champaign Illinois: AOCS p. 30-39.

Picot L, Bordenave S, Didelot S, Fruitier-Arnaudin I, Sannier F, Thorkelsson G, Bergé JP, Guérard F, Chabeaud A, Piot JM. 2006. Antiproliferative activity of fish protein hydrolysates on human breast cancer cell lines. Process Biochem 41:1217-1222.

Pittas AG, Lau J, Hu FB, Dawson-Hughes B. 2007. The role of vitamin D and calcium in type 2 diabetes. A systematic review and meta-analysis. J Clin Endocrinol Metab 92(6):2017-2029.

Ponder DL, Innis SM, Benson JD, Siegman JS. 1992. Docosahexaenoic acid status of term infants fed breast milk or infant formula containing soy oil or corn oil. Pediatric Research 32:421-426.

Poole EM, Bigler J, Whitton J, Sibert JG, Kulmacz JR, Potter JD, Ulrich CM. 2007. Genetic variability in prostaglandin synthesis, fish intake and risk of colorectal polyps. Carcinogenesis 28(6):1259-1263.

Psota TL, Gebauer SK, Kris-Etherton P, 2006. Dietary omega-3 fatty acid intake and cardiovascular risk. Am J Cardiol 98(4A):3i-18i.

Quijano MA, Moreno P, Gutiérrez AM, Pérez-Conde MC, Cámara C. 2000. Selenium speciation in animal tissues after enzymatic digestion by high-performance liquid chromatography coupled to inductively coupled plasma mass spectrometry. J Mass Spectrom 35:878-884.

Ramel A, Jonsdottir MT, Thorsdottir I. 2008. Consumption of lean fish and weight loss in young overweight and obese adults on an energy reduced diet for eight weeks. Nutrition, Metabolism and cardiovascular diseases. American Journal of Clinical Nutrition 87(4):1045-1052.

Rampone AJ, Machida CM. 1981. Mode of action of lecithin in suppressing cholesterol absorption. J Lipid Res 22:744-752.

Rayman MP. 2002. The argument for increasing selenium intake. Proceedings of the Nutrition Society 61:203-215.

Rayman MP. 2005. Selenium in cancer prevention: a review of the evidence and mechanism of action. Proceedings of the Nutrition Society 64:527-542.

Richy F, Bruyere O, Ethgen O, Cucherat M, Henrotin Y, Regenster J. 2003. Structural and symtomatic efficiacy of glucosamine and chondroitin in knee osteoarthritis: A comprehensive meta-analysis. Archives of Internal Medicine 163:1514-1522.

Robinson JG, Stone NJ. 2006 Antiatherosclerotic and antithrombotic effects of omega-3 fatty acids. Am J Cardiol 21:98(4A):39i-49i.

Roe DA, Weston MO. 1965. Potential Significance of Free Taurine in the Diet. Nature 205:287-288.

Romieu I, Torrent M, Garcia-Esteban R, Ferrer C, Ribas-Fitó N, Antó JM, Sunyer J. 2007. Maternal fish intake during pregnancy and atopy and asthma in infancy. Clin Exp Allergy 37(4):518-525.

Rose WC. 1976. Amino-Acid Requirements of Man. Nutrition Reviews 34(10):307-309.

Ruel-Gariepy E, Leroux J-C. 2006. Chitosan: A natural polycation with multiple applications. Pharmaceutical Appl. ACS Symp Ser 934:243-259.

Salam MT, Li YF, Langholz B, Gilliland FD. 2005. Maternal fish consumption during pregnancy and risk of early childhood asthma. J Asthma 42(6):513-518.

Sala-Vila A, Castellote A, Campoy C, Rivero M, Rodriguez-Palmero M, Lopez-Sabater MC. 2004. The Source of Long-Chain PUFA in Formula Supplements Does Not Affect the Fatty Acid Composition of Plasma Lipids in Full-Term Infants. Journal of Nutrition 134(4):868-873.

Saldeen T, Wallin R, Marklinder I. 1998. Effects of a small dose of stable fish oil substituted for margarine in bread on plasma phospholipid fatty acids and serum triglycerides. Nutr Res 18:1483-1492.

Salvatore S, Heuschkel R, Tomlin S, Davies SE, Edwards S, Walker-Smith JA, French I, Murch SH. 2000. A pilot study of N-acetyl glucosamine, a nutritional substrate for glycosaminoglycan synthesis, in paediatric chronic inflammatory bowel disease. Aliment Pharmacol Therapeutics 14:1567-1579.

Sanders TAB. 2000. Polyunsaturated fatty acids in the food chain in Europe. Am J Clin Nutr 71:176S-178S.

Sathivel S. 2003. Biochemical and functional properties of herring (*Clupea harengus*) byproduct hydrolysates. J Food Sci 68:2196-2200.

Sausenthaler S, Koletzko S, Schaaf B, Lehmann I, Borte M, HerbarthO, von Berg A, Wichmann H-E, Heinrich J. 2007. Maternal diet during pregnancy in relation to eczema and allergic sensitization in the offspring at 2 y of age. Am J Clin Nutr 85(2):530-537.

Schaafsma, G. 2007. The Role of Nutrition in Joint Health Promotion. Nutrafoods Research 6(3):5-11.

Scientific Committee for Food, Commission of the European Communities. 1993. Thirty-first series, Nutrient and energy intakes for the European Community. Reports of the Scientific Committee for Food of the European Community, Commission of the European Communities. Luxembourg, p. 1-255. http://europa.eu.int/comm/food/fs/sc/scf/out89.pdf

Scientific Committee for Food, Commission of the European Communities. 2000. Opinion of the Scientific Committee on Food on the Tolerable Upper Intake Level of Selenium, SCF/CS/NUT/UPPLEV/25 Final, p. 1-18.

Schmelz EM. 2003. Dietary sphingolipids in the prevention and treatment of colon cancer. In Szuhaj BW, van Nieuwenhuyzen W (eds.) Nutrition and biochemistry of phospholipids. Champaign Illinois: AOCS Press p. 80-87.

Seierstad SL, Seljeflot I, Johansen O, Hansen R, Haugen M, Rosenlund G, Frøyland L, Arnesen H. 2005. Dietary intake of differently fed salmon; the influence on markers of human atherosclerosis. Eur J Clin Invest 35(1):52-59.

Shao A, Hathcock JN. 2008. Risk assessment for the amino acids taurine, L-glutamine and L-arginine. Regulatory Toxicology and Pharmacology 50:376-399.

Shen L, Hoek-van Nieuwenhuizen M, Luten JB. 1997. Speciation and *in vitro* bioavailability of selenium in fishery products. In: Luten JB, Børresen T, Oehlenschläger J (eds.) Seafood from Producer to Consumer, Integrated Approach to Quality, Amsterdam: Elsevier p. 653-663.

Siezen CL, Bas Bueno-de-Mesquita H, Peeters PHM, Kram NR, van Doeselaar M, van Kranen HJ. 2006. Polymorphisms in the genes involved in the arachidonic acid-pathway, fish consumption and the risk of colorectal cancer. International Journal of Cancer 119(2):297-303.

Silvers KM, Scott KM. 2002. Fish consumption and self-reported physical and mental health status. Public Health Nutrition 5(3):427-431.

Simmer K, Patole S. 2004. Longchain polyunsaturated fatty acid supplementation in preterm infants. Cochrane Database Syst Rev 2004:CD000375.

Simmer K. 2004. Longchain polyunsaturated fatty acid supplementation in infants born at term (Cochrane Review). Cochrane Database Syst Rev 2004:1-25.

Simopoulos AP, Salem, N. Jr. 1992. Egg yolk as a source of long-chain polyunsaturated fatty acids in infant feeding. Americal Journal of Clinical Nutrition 55:411-414.

Simopoulos AP. 2006. Evolutionary aspects of diet, the omega-6/omega-3 ratio and genetic variation: nutritional implications for chronic diseases. Biomed Pharmacother 60(9):502-507.

Siwek M, Galunsky B, Niemeyer B. 2005. Isolation of selenium organic species from antarctic krill after enzymatic hydrolysis, Anal Bioanal Chem 381:737-741.

Skerrett PJ, Hennekens CH. 2003. Consumption of fish and fish oils and decreased risk of stroke. Prev Cardiol 6(1):38-41.

Song JH, Inoue Y, Miyazawa T. 1997. Oxidative stability of docosahexaenoic acid-containing oils in the form of phospholipids, triacylglycerols, and ethyl esters, Bioscience, biotechnology, and biochemistry 61(12):2085-2088.

Soucy J, LeBlanc J. 1998. Effects of beef steak and cod fillet on plasma glucose insulin, amino acids and energy metabolism in normal subjects. Nutrition Research 18(7): 1113-1123.

Specker B. 2004. Vitamin D requirements during pregnancy. Am J Clin Nutr 80(6 Suppl):1740S-1747S.

Spitze AR, Wong DL, Rogers QR, Fascetti AJ. 2003. Taurine concentrations in animal feed ingredients; cooking influences taurine content. Journal of Animal Physiology and Animal Nutrition 87(7-8):251-262.

Stalenhoef AF, de Graaf J, Wittekoek ME, Bredie SJ, Demacker PN, Kastelein JJ. 2000. The effect of concentrated n-3 fatty acids versus gemfibrozil on plasma lipiproteins, low density lipoprotein heterogeneity and oxidizability in patients with hypertriglyceridemia. Atherosclerosis 153:129-138.

Stapleton PP, Charles R, Redmond H, Bouchier-Hayes D. 1997. Taurine and human nutrition. Clinical Nutrition 16(3):103-108.

Stripp C, Overvad K, Christensen J, Thomsen BL, Olsen A, Møller S, Tjønneland A. 2003. Fish intake is positively associated with breast cancer incidence rate. J Nutr 133(11):3664-3669.

Szajewska H, Horvath A, Koletzko B. 2006. Effect of n-3 long-chain polyunsaturated fatty acid supplementation of women with low-risk pregnancies on pregnancy outcomes and growth measures at birth: a meta-analysis of randomized controlled trials. Am J Clin Nutr 83(6):1337-1344.

Szuhaj BF, Van Nieuwenhuyzen W. (eds.) In: Nutrition and biochemistry of phospholipids. Champaign Illinois: AOCS Press iii 242 pp.

Tanskanen A, Hibbeln JR, Tuomilehto J, Uutela A, Haukkala A, Viinamäki H, Lehtonen J, Vartiainen E. 2001. Fish consumption and depressive symptoms in the general population in Finland. Psychiatric Services 52(4):529-531.

Tavani A, Franceschi S, Levi F, La Vecchia C. 2005. Fish, omega-3 polyunsaturated fat intake and cancer at selected sites. World Rev Nutr Diet 94:166-175.

Terry PD, Rohan TE, Wolk A. 2003. Intakes of fish and marine fatty acids and the risks of cancers of the breast and prostate and of other hormone-related cancers: a review of the epidemiologic evidence. Am J Clin Nutr 77(3):532-543.

Thonthtai C, Gildberg A. 2005. Asian fish sauce as a source of nutrition. In: Shi J, Ho C-T, Shahidi F (eds.) London: Asian Functional Foods CRC Press p. 215-265.

Thorngren M. 1983. Dietary fish and haemostasis, Indications of anti-thrombotic properties, in Department of internal medicine and community health science. Lund: Lund Universityal hospital.

Thorsdottir I, Tomasson H, Gunnarsdottir I, Gisladottir E, Kiely M, Parra MD, Bandarra NM, Schaafsma G, Martinéz JA. 2007. Randomized trial of weight-loss-diets for young adults varying in fish and fish oil content. Int J Obes (Lond) 31(10):1560-1566.

Timonen M, Horrobin D, Jokelainen J, Laitinen J, Herva A, Räsänen P. 2004. Fish consumption and depression: the Northern Finland 1966 birth cohort study. Journal of Affective Disorders 82(3):447-452.

Tremblay F, Lavigne C, Jacques H, Marette A. 2003. Dietary cod protein restores insulin-induced activation of phoshadtidylinositol 3-kinase/Akt and GLUT4 translocation in the T-tubules in skeletal muscle of High-fat-fed obese rats. Diabetes 52:29-37.

Tremblay F, Jacques H, Marette A. 2005. Modulation of insulin action by dietary proteins and amino acids: role of the mammalian target of of rapamycin nutrient sensing pathway. Current Opinion in Clinical Nutrition and Metabolic Care 8:457-462.

Tremblay F, Lavigne C, Jacques H, Marette A. 2007. Role of dietary proteins in the pathogenesis of insulin resistance. Annual Review of Nutrition 27:293-310.

Uhe AM, Collier GR, Odea K. 1992. A Comparison of the Effects of Beef, Chicken and Fish-Protein on Satiety and Amino-Acid Profiles in Lean Male-Subjects. Journal of Nutrition 122(3):467-472.

Undeland I, Ellegard L, Sandberg A-S. 2004. Fish and cardiovascular health. Scandinavian Journal of Nutrition 48(3):119-130.

Van de Poll MC, Dejong CH, Soeters PB. 2006. Adequate range for sulfur-containing amino acids and biomarkers for their excess: lessons from enteral and parenteral nutrition. J Nutr 136(6 Suppl):1694S-1700S.

Virtanen JK, Rissanen T, Voutilainen S, Tuomainen T. 2007. Mercury is a risk factor for cardiovascular diseases. Journal of Nutritional Biochemistry 18:75-85.

Visioli F, Rise P, Plasmati E, Pazzucconi F, Sirtori CR, Galli C. 2000. Very low intakes of N-3 fatty acids incorporated into bovine milk reduce plasma triacylglycerol and increase HDL-cholesterol concentrations in healthy subjects. Pharmacol Res 41:571-576.

Wallace JMW, McCabe AJ, Robson PJ, Keogh MK, Murray CA, Kelly PM, Marquez-Ruiz G, McGlynn H, Gilmora WS, Strain JJ. 2000. Bioavailability of n-3 polyunsaturated fatty acids (PUFA) in foods enriched with microencapsulated fish oil. Ann Nutr Metab 44:157-162.

Wang C, Harris WS, Chung M, Lichtenstein AH, Balk EM, Kupelnick B, Jordan HS, Lau J. 2006. n-3 Fatty acids from fish or fish-oil supplements, but not alpha-linolenic acid, benefit cardiovascular disease outcomes in primary- and secondary-prevention studies: a systematic review. Am J Clin Nutr 84(1):5-17.

Watson KE, Abrolat ML, Malone LL, Hoeg JM, Doherty T, Detrano R, Demer LL. 1997. Active serum vitamin D levels are inversely correlated with coronary calcification. Circulation 96(6):1755-1760.

Werner A, Havinga R, Kuipers F, Verkade HJ. 2004. Treatment of EFA deficiency with dietary triglycerides or phospholipids in a murine model of extrahepatic cholestasis. American Journal of Physiology, Gastrointestinal and Liver Physiology 286:822-826.

Whanger PD. 2004. Selenium and its relationship to cancer: an update. British Journal of Nutrition. 91(1):11-28.

Whelton SP, He J, Whelton PK, Muntner P. 2004. Meta-analysis of observational studies on fish intake and coronary heart disease. Am J Cardiol 93(9):1119-1123.

Wijendran VASU, Huang MC, Diau GY, Boehm GUNT, Nathanielsz PW, Brenna JT. 2002. Efficacy of Dietary Arachidonic Acid Provided as Triglyceride or Phospholipid as Substrates for Brain Arachidonic Acid Accretion in Baboon Neonates. Pediatric Research 51(3):265-272.

Wijendran V, Hayes KC. 2004. Dietary n-6 and n-3 fatty acid balance and cardiovascular health. Annu Rev Nutr 24:597-615.

Willers S, Devereux, G, Craig, LCA, McNeill G, Wijga AH, El-Magd W Abou, Turner SW, Helms PJ, Seaton A. 2007. Maternal food consumption during pregnancy and asthma, respiratory and atopic symptoms in 5-year-old children. Thorax 27:27.

World Cancer Research Fund, Food, Nutrition, Physical Activity, and the Prevention of Cancer: a Global Perspective. 2007. Washington DC: American Institute for Cancer Research.

Xiong S-L, An A-L, Jin Z-Y, Chen M. 2007. Effects of oral chondroitin sulphate on lipid and antioxidant metabolisms in rats fed a high-fat diet. Journal of Food Biochemistry 31(1):356-369.

Xiong S-L, Jin Z-Y. 2007. The free radical-scavenging property of chondroitin sulphate from pig laryngeal cartilage *in vitro*. Journal of Food Biochemistry 31(1):28-44.

Yamada T, Strong JP, Ishii T, Ueno T, Koyama M, Wagayama H, Shimizu A, Sakai T, Malcom GT, Guzman MA. 2000. Atherosclerosis and -3 fatty acids in the populations of a fishing village and a farming village in Japan. Atherosclerosis 153:469-481.

Yamori Y, Nara Y, Mizushima S, Sawamura M, Horie R. 1994. Nutritional factors for stroke and major cardiovascular diseases: international epidemiological comparison of dietary prevention. Health Rep 6(1):22-27.

Yamori Y, Liu L, Ikeda K, Miura A, Mizushima S, Miki T, Nara Y. 2001. Distribution of twenty-four hour urinary taurine excretion and association with ischemic heart disease mortality in 24 populations of 16 countries: Results from the WHO-CARDIAC Study. Hypertension Research 24(4):453-457.

Yamori Y, Murakami S, Ikeda K, Nara Y. 2004. Fish and life style disease prevention: Experimental and epidemiological evidence for anti-atherogenic potential of taurine. Clin Exp Pharmacol Physiol. 31 Suppl 2:S20-3.

Yamori Y, Liu L, Mizushima S, Ikeda K, Nara Y. 2006. Male cardiovascular mortality and dietary markers in 25 population samples of 16 countries. J Hypertens 24(8):1499-1505.

Yang CX, Takezaki T, Hirose K, Inoue M, Huang XE, Tajima K. 2003. Fish consumption and colorectal cancer: a case-reference study in Japan. European Journal Of Cancer Prevention 12(2):109-115.

Yuan JM, Ross RK, Gao YT, Yu MC. 2001. Fish and shellfish consumption in relation to death from myocardial infarction among men in Shanghai, China. American Journal of Epidemiology 154(9):809-816.

Zampelas A, Panagiotakos D, Pitsavos C, Das U, Chrysohoou C, Skoumas Y, Stefanadis C. 2005. Fish consumption among healthy adults is associated with decreased levels of inflammatory markers related to cardiovascular disease: the ATTICA study. J Am Coll Cardiol 46(1):120-124.

Zhang J, Sasaki S, Amano K, Kesteloot H. 1999. Fish consumption and mortality from all causes, ischemic heart disease, and stroke: an ecological study. Prev Med 28(5):520-529.

Zhao X, Jia J, Lin Y. 1998. Taurine content in Chinese food and daily intake of Chinese men. Adv Exp Med Biol 442:501-505.

Zittermann A, Schleithoff SS, Koerfer R. 2007. Vitamin D and vascular calcification. Curr Opin Lipidol 18(1):41-46.

Zittermann A. 2006. Vitamin D and disease prevention with special reference to cardiovascular disease. Prog Biophys Mol Biol 92(1):39-48.

Processing of marine lipids and factors affecting their quality when used for functional foods

***Charlotte Jacobsen*[1]*, Turid Rustad*[2]*, Nina Skall Nielsen*[1]*, Eva Falch*[3]*, Stig Jansson*[4] *and Ivar Storrø*[5]**

[1]*National Institute of Aquatic Resources, Technical University of Denmark, 2800 Kgs. Lyngby, Denmark*

[2]*The Norwegian University of Science and Technology, Department of Biotechnology, Trondheim, Norway*

[3]*03C Nutraceuticals AS, Strandgata 84, 6512 Kristiansund, Norway*

[4]*Noromega AS, Sjøgata 15/17, 8006 Bodø, Norway*

[5]*SINTEF Fisheries and Aquaculture, Trondheim, Norway*

1. General introduction

The use of marine lipids as ingredients in functional foods is increasing due to their well- documented health benefits. When applying marine lipids in foods it is of the utmost importance that they have a high quality so that the sensory and functional properties and shelf-life of the final product are acceptable.

This review has been written to create an overview of the methods used for processing marine lipids as well as of the different factors that can affect the quality of marine lipids used as ingredients in functional foods. Quality parameters include level of healthy omega-3 fatty acids, oxidative stability and levels of contaminants. The review deals with the factors that can affect the quality of the raw material as well as factors influencing the quality during processing and the final application of the lipids. Moreover, the review also contains general information about the raw material to be used for the production of marine lipid-based ingredients for functional food products.

2. Quality parameters

When dealing with marine lipids some of the most important quality aspects are: (1) the level of long chain omega-3 polyunsaturated fatty acids (PUFA); (2) the level of different lipid classes (phospholipids, triglycerides, free fatty acids and other lipids such as squalene); (3) contamination by lipophilic compounds such as dioxin, PCB and other environmental pollutants; and (4) lipid oxidation. This chapter will deal with all these aspects. As an introduction the lipid oxidation processes termed autooxidation and photo-oxidation will briefly be described. A simplified description of the lipid oxidation process is presented in Figure 1.

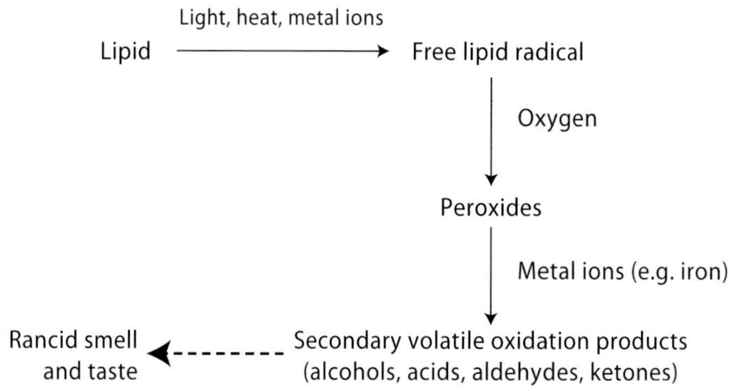

Figure 1. *Simplified description of lipid oxidation reactions.*

The lipid oxidation process consists of the following steps. Initially, the polyunsaturated lipid will, in the presence of coloured sensitizers and light, metal ions or heat, form free radicals. These free lipid radicals will react with oxygen whereby peroxyl radicals are formed. Subsequently, the peroxyl radical will react with a new lipid molecule whereby the lipid hydroperoxides (peroxides) are formed. The level of peroxides is often measured by the peroxide value (PV). The peroxides are also termed primary oxidation products and they are tasteless. However, peroxides may in the presence of heat or metal ions be decomposed into secondary volatile oxidation products (volatiles), which are responsible for the off-flavours formed due to oxidation (often described as rancid, fishy, etc.). The level of volatiles can be measured by different spectrophotometric methods (e.g. anisidine value and thiobarbituric acid reactive substances (TBARS)) or by headspace gas chromatographic techniques.

3. Factors important for raw material quality

To obtain marine lipids of high quality for functional food applications selection of high quality raw material is important. Selection, storage and handling of the raw material are preliminary steps in the processing of marine lipids. These steps depend both on the needs and demands of end users as well as the needs and wishes of the raw material producers. The fish that are processed to produce crude fish oil (and fish meal) can usually be categorised as follows: (1) co-products from the edible fisheries, e.g. cutting from filleting industry; (2) parts of the fish that are not usually used for human consumption; or (3) fish types that are not considered acceptable or aesthetically pleasing for human consumption. The latter are caught especially for reduction to fish meal and fish oil. The fatty acid composition of fish oil depends on the fatty acid composition of the feed and therefore substantial variation is observed within each species.

Co-products come both from fisheries and aquaculture. These products must be collected, disposed of and/or utilised in the most cost-effective manner. For both co-products and whole fish, spoilage and the development of off-flavour must be prevented by controlling lipid oxidation and protein degradation. Raw material with amounts of chemical contaminants like PCBs and dioxins exceeding maximum limits and action limits set by authorities must be avoided (Recommendation 2004/705/EC).

The factors that can affect the quality of the raw material are summarised in the following.

3.1 Environment

The geographic region of catching influences contaminants building up in the food chain and determines the final content of persistent halogenated compounds like dioxin, dioxin-like PCBs as well as trace elements such as lead, copper, cadmium, arsenic, caesium and many other contaminants. Other important factors to consider are the marine animal's adaption to the environment in general, such as sea temperature that varies between -2 °C versus 30 °C. Seasonal variations have an influence on food availability, spawning and also geographic migration, which can all have a major influence on the raw material quality and composition.

3.2 Species and seasonal variations

Species dependent factors like position in the food chain result in variation in the accumulation of environmental pollutants. Survival strategy of the species varies and results in the accumulation of different amounts and composition of lipids as well as place of deposition. The survival strategy also has an impact on pollutants accumulated in the specific species, i.e. short life-cycle compared to long life-cycle. The strategy of deposition can also have an impact on oil quality since lipids and contaminants in the organ selected for extraction can vary.

Lean or fatty fish refers only to the muscle content of lipids and does not take into account that the fish can deposit fat as energy in the organs as well. The location of the lipid for energy storage in fish varies widely between species. Fish that lack a swim bladder tend to accumulate less dense lipids like hydrocarbons as squalene to increase buoyancy. Cod fishes and shark family deposit lipids in the liver and salmon, mackerel and herring store lipids in the muscle, whereas anchovy stores the depot lipids subcutaneously.

The food chain starts with phytoplankton synthesising the long chain omega-3 polyunsaturated fatty acids with 5 and 6 double bonds (EPA and DHA) that accumulate in the food chain (Gunstone *et al.* 1994). The fatty acids of triacylglycerols in depot lipids are strongly dependent on the diet (Shirai *et al.* 2002).

3.3 Lipid oxidation in raw material

Lipid oxidation in fish muscle can be caused by non-enzymatic processes such as autooxidation and photosensitised oxidation, as well as catalysed by enzymes such as lipoxygenase (Decker *et al.* 1988). Lipoxygenase activity has been detected in various tissues of fishes and shrimp in blood plasma, gill, skin, fish eggs, brain, muscle, erythrocytes and platelets (Pan and Kuo 2000). In general the susceptibility towards lipid oxidation increase in the following order; white muscle, dark muscle and skin (Undeland *et al.* 1998, 1999; Sikorski and Kolakowski 2000). The dark muscle (slow muscle) is dense in mitochondria organelles that utilise lipids (TAG) to generate energy through the aerobic pathway by β-oxidation. The dark muscle contains high concentrations of prooxidative haemoproteins like haemoglobin and its derivatives. The mitochondria organelles are dense in membranes, and are therefore rich in phospholipids and cholesterol, and a higher concentration of phospholipids can be found in the dark muscle versus white muscle (Ingemansson *et al.* 1991)

Metal ions such as Fe are in general able to catalyse lipid oxidation. In fish muscle, several studies seem to suggest that heme Fe is a stronger prooxidant than low molecular weight Fe (Chiu *et al.* 1996; Undeland *et al.* 2002). Fish blood can induce increased lipid oxidation and bled fish tend to develop less rancidity when processed to mince (Richards and Hultin 2002). Both haemoglobin and myoglobin can act as prooxidants in meat products. At pH values of relevance for meat and meat products (pH 5.5), both metmyoglobin and oxymyoglobin have been shown to be major initiators of lipid oxidation. Oxidation induced by the presence of myoglobin is highly pH dependent. Surprisingly, much of the data obtained in simple model systems regarding myoglobin induced lipid oxidation is compatible with empirical data observed *in vivo*, *ex vivo*, and in muscle foods (Baron and Andersen 2002).

Antioxidant acting enzymes in fish, superoxide dismutase, catalase, glutathione peroxidase, and glutathione reductase have protective roles against lipid oxidation and their concentration varies between species and organs (Trenzado *et al.* 2006). Antioxidants common in fish are tocopherol, astaxanthin, ubiquinol and vitamin C. Tocopherol show an inverse relationship with total lipid content, i.e. higher concentrations in lipid from lean fish versus lipid rich fish (Gunstone *et al.* 1994). *Post mortem* storage on ice or freezing will lead to depletion of antioxidants due to oxidation (Ingemansson *et al.* 1993; Passi *et al.* 2005; Undeland *et al.* 1999). Results obtained with fish stored on ice 48 hours *post mortem,* showed that vitamin C and ubiquinol was almost completely depleted. The distribution of the antioxidants differs between organs (Elvevoll 1988; Ingemansson *et al.* 1993; Passi *et al.* 2005). Some antioxidants are highly sensitive to partial oxygen pressure and the antioxidative effect *post mortem* can therefore disappear when diffusion of oxygen and partial pressure increase (Burton and Ingold 1984; Ingemansson *et al.* 1993; Passi *et al.* 2005).

Proteins may also undergo oxidative deterioration. Oxidised proteins may interact with fresh and oxidised lipids and this may influence the oxidative stability of the lipids. Oxidation can occur at both the protein backbone and on the amino acid side chains, with the ratio of attack depending on a number of factors (Davies 2005). Oxidation can result in major physical changes in protein structure ranging from fragmentation of the backbone to oxidation of the side-chains. Oxidation of protein side-chains can give rise to unfolding and conformational changes in protein and also to dimerisation or aggregation (Davies 2005).

Oxidation of both the backbone and the side-chains can result in the formation of further reactive species: such as hydroperoxides or peroxides, and other short-lived intermediates (Davies 2005). The most important reactive oxygen species are hydroxyl radical and singlet oxygen. Hydrogen peroxide and superoxide anion are important precursors for hydroxyl radical and singlet oxygen formation.

Lipid - protein interaction was found to depend on oxygen level, formation of lipid oxidation products (Aubourg and Medina 1999), storage temperature and time (Aubourg and Medina 1999; Aubourg *et al.* 1998), degree of unsaturation of lipid (Liu and Wang 2005), contact area between lipid and protein (Rampon *et al.* 2001), concentration (Funes *et al.* 1982), amount of prooxidants and antioxidants.

3.4 Lipolytic activity

Between harvest and processing into fish oil there is a rapid change in the properties/quality of fish. This is due to enzymatic processes caused by endogenous (autolytic process) and bacterial enzymes. The rates of these processes are highly temperature dependent and chilling and freezing reduce the reaction rate. The lysosomes contain numerous enzymes capable of degrading both carbohydrates, lipids, nucleic acids and proteins.

After death, lipases and phospholipases present in the fish muscle will hydrolyse lipids producing free fatty acids and partially hydrolyzed lipids (diglycerides and monoglycerides), glycerol and nitrogen bases. This wide variety of relatively small molecules can undergo further interactions with fish constituents leading to quality losses in the product. Lipases and phospholipases are still active during frozen storage. As a result, a wide range of lipid hydrolysis compounds are produced, most of them relatively small molecules that are susceptible to interaction with fish constituents leading to a lowering in nutritional and sensory values of the product (Auborg 2001). Lipolytic enzymes in fish are responsible for digesting lipids in intestines (bile salt stimulated lipase and pancreas lipase), rebuilding lipids for transport in lipoprotein vehicles (lipoprotein lipase) and for further hydrolysis to pass cell membrane and enter cell interior (hepatic lipase), for re-esterification to store lipids in adipose tissue or to be utilized as energy. From adipose tissue the lipids are further released by intracellular lipase ('hormone' stimulated lipase) again and transported out of the cell and distributed by lipoproteins to energy demanding cells requiring fuel in mitochondrial oxidation (Sheridan 1988). Besides lipase originating from fish tissue cells, lipolytic activity in the intestinal region that possible can originate from bacterial lipases naturally occurring in the intestinal region has also been reported (Lopez-Amaya and Marangoni 2000).

A recent study showed that lipase activity is depleted at temperatures in the region of 60-65°C in both viscera, cut-off and liver of cod (Sovik and Rustad 2005). Investigation of activity at pH 5 and 7 showed that liver and cut-off had highest activity at acid condition and opposite for viscera. The lipase activity in the viscera was bile salt activated. Phospholipase A (not specified) from pollack muscle tissue had optimum activity between 37-42 °C at a pH of 8.5-9.0 and showed Ca^{2+} dependency and inhibition by EDTA (Aaudley *et al.* 1978). From bonito muscle a non Ca^{2+} dependent phospholipase A_1 has been isolated. This enzyme showed the highest activity at pH 6.5-7.0 and 20-30 °C. Phospholipase A_2 from liver of rainbow trout has also been investigated for temperature dependent deacylation at physiological temperatures (Neas and Hazel 1984). Little has been found published on heat inactivation or stability of phospholipases in fish. Phospholipase A_2 from cod muscle showed acid activity with optimum at pH 4. The activity was found to be independent of Ca^{2+} and a temperature of 55 °C resulted in 50% reduction in activity compared to optimum temperature 40 °C (Aaen *et al.* 1995).

4. Processing of marine lipids

4.1 Fish meal processing

Rendering into dried meal and crude oil is the most usual way of processing small oily fish and trimmings from processing fish for food. During the processing fish oil is produced by pressing it out of the cooked fish. The process is described on the home page of the International Fishmeal and Fish Oil Organization (http://www.gafta.com/fin/) and in a technical paper from the FAO (2004) and is summarised below. The annual global production of fish body oil is about 1.2 million tons. It comes mostly from about 400 fish meal plants. South America usually provides the bulk but the Far East and South East Asia (12%) are also major producers. In Europe, Denmark, Iceland and Norway are all significant suppliers, each providing around 5% of the global supply. The South American supply mostly consists of anchovy whereas capelin and sand eel are the main constituents of European supplies (GRP 2007).

4.1.1 Extraction technology

The lipids are embedded in tissue and it is necessary to burst or destroy the cell membrane to liberate the oil. The extraction differs in method and technology to burst cells and in media that are used to facilitate the extraction. Some of the methods that are utilised are heat denaturation, chemical hydrolysis using acids or bases, denaturation by pH, osmotic, enzyme hydrolysis, mechanical pressure and shear forces.

In all cases it is important to grind the raw material to increase the liberation of oil to maximise yield. To isolate the liberated oil the most common way is to make use of mechanically induced gravity by centrifugal separation in 2- or 3-phase separators, 2-phase decanter as well as tricanter (3-phase). Supercritical extraction may also be applied.

All methods to facilitate extraction have limitations depending on what is crucial for product quality, target compound to be isolated in end product, down stream processing and safety at production location (off shore on shore, etc.). The methods are further discussed below.

4.1.1.1 Heat denaturation

Wet and dry extraction differs in that wet extraction uses direct live steam injection into the raw material. This method is naturally fading out in industry. Dry extraction refers to heating with steam jacket and is the most common method. Some methods add water and grind the material to easier liberate relieved oil and to overcome emulsification problems.

Other methods used are the Friolex method by Westfalia, where water is mixed with alcohol (ethanol) to assist the extraction. Different salts may also be added (Hruschka and Frische 1998). This can allow for an immediately removal of free fatty acids (FFA) simultaneously as oil is separated in the centrifuge.

4.1.1.2 Enzymatic hydrolysis

Enzyme assisted liberation of oil has become popular and has proven to work. The extended time of reaction can be from 30 min and up to normally no more than 60 min. The process normally uses less costly industrial enzymes like alcalase or similar products. At the end of the process the enzyme is heat inactivated and the heat will also facilitate a separation. Another procedure is to separate before inactivation (Gbogouri *et al.* 2006).

4.1.1.3 Cold pressing

Cold pressing refers to methods of squeezing and pressing facilitated by a screw press or similar device at low temperatures as for olive oil extraction. Difficulties arise since the material is viscoelastic and mechanical forces are aligned throughout the tissue. This method is therefore not common due to poor yields.

4.1.1.4 Low temperature

However, if combined with the freezing of raw material, the intra- and extra cellular supported networks of stretching structure that normally align mechanical forces are ruled out. Mechanical grinding, or forces in general on the raw material, easily disrupt the cell membrane so that the oil may be extracted by moderate heating by melting only the crystallised oil (Jansson and Elvevoll 2000).

Another method described for extracting fish oils uses excessive amounts of cold water that mixes with the grinded raw material and is processed under mild conditions (Barrier and Rousseau 1998).

4.1.1.5 Solvent extraction

Hexane and acetone are among solvents that can be used, but these methods are not commonly used.

4.1.1.6 Supercritical extraction

Supercritical extraction with CO_2, propanol or other appropriate supercritical media can also be used for lipid extraction.

Very fresh raw materials from aquaculture can give oils of high quality but the oil might be of variable stability depending on the exact composition of the oil and of the method to extract the oil. There are several producers making fish oil within hours after fish slaughter.

4.2 Refining of crude fish oil

After extraction of the oil from the fish, the crude fish oil is refined to remove impurities that cause the original product to have an unattractive colour or taste or that cause harmful metabolic effects (Bimbo 2007). These impurities include compounds formed as a result of lipid oxidation. At the same time, processing should retain desirable nutritional components (e.g. omega-3 PUFA and tocopherol). As mentioned earlier the challenge is that the threshold of volatile compounds in marine oils can easily be detected in ppb levels. Therefore, to obtain satisfying results for refining high stability marine oils, high quality crude oil is required. Otherwise the cleaning of the oil will be too vigorous and harsh and result in a declining natural stability which can hardly or never be recovered. The dioxin scandal in Belgium in 1999 resulted in a demanding task for the marine oil producers who had to cope with the new legalisation on this issue.

The key is to maintain the naturally present antioxidants while removing prooxidants like transition metals, taste and odour compounds that are secondary oxidation products such as aldehydes and ketones among others. Environmental pollutants also have to be removed and especially those of a halogenated nature like PCBs, dioxin, furans and others that are included in the index of WHO TEQ. Bromated flame retardants are an up-and-coming issue, but not yet established in legalisation and will also be of interest to remove. As analytic methods as well as technology for the removal of contaminants improve, the governmental authorities will push the limits allowable and will lower the limits in 2008 (European Commission 2006). For the moment marine oil for human consumption has the following limit: sum of dioxins and furans (WHO-PCDD/F-TEQ) 2.0 pg/g lipid and sum of dioxins, furans and dioxin-like PCBs (WHO-PCDD/F-PCB-TEQ) 10.0 pg/g lipid.

Increasing knowledge about the necessity of customised technology and processing have further improved the quality in general, but most of the technology and performance used for marine oils, rely on the transfer of experience from far less sensitive oils in the field vegetable oil industry.

The different possible steps in the refining process are described below (see also Figure 2) with the emphasis on the compounds that are removed during the different steps.

4.2.1 Degumming

Degumming by treatment of the marine oil with phosphoric acid or other acids removes phospholipids, proteinaceous compounds and trace metals. The removal of phospholipids and trace metals will improve the oxidative stability (Bimbo 2007).

4.2.2 Neutralisation

Neutralisation by the addition of an alkali solution removes free fatty acids, pigments, phospholipids, oil insolubles, water solubles and trace metals (Bimbo 1990). The reduction in the content of free fatty acids will improve the sensory properties and

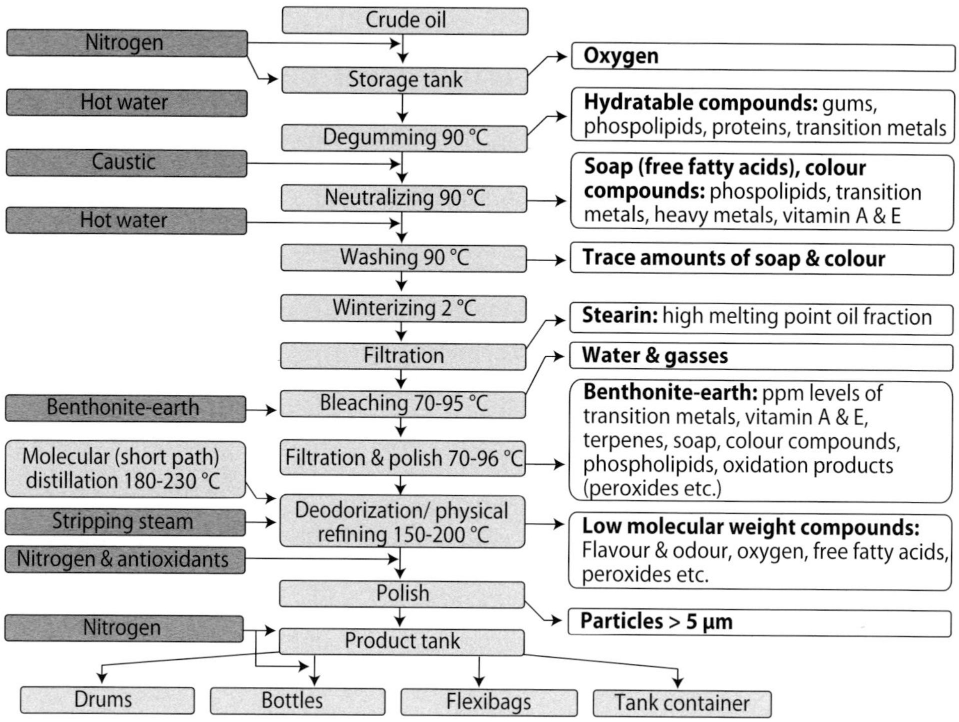

Figure 2. The refining and deodorisation process.

oxidative stability of the oil. The free fatty acid content of refined fish oils should be as low as possible, preferably not higher than 0.1-0.2%.

Improvements have been made where synthetic silica hydrogels are used to reduce soap levels so washing in alkali refining can be eliminated. This process has been improved and resulted in the development of Modified Caustic Refining (Welsh *et al.* 1989) and Modified Physical Refining (MPR) (Toeneboehn *et al.* 1994). In MPR the adsorption capability of silica is necessary to remove phospholipids and trace amounts of transition metals, which is a crucial step to obtain a stable process and oil. Magnesium silica has shown effective capture of oxidised lipids and products thereof (Yates and Caldwell 2000). The use of silica to process vegetable oils and marine oils (Hernandez and Rathbone 2002; Silkeberg *et al.* 1998) has shown less adsorption of natural antioxidants, which improves stability in the oil itself.

4.2.3 Bleaching

Bleaching involves the adsorption of coloured compounds; peroxides and some volatile oxidation compounds as well as other impurities to the bleaching clay and this will improve the oxidative stability (Bimbo 2007).

In traditional bleaching, clays used are often of an activated nature. This in combination with the rest of the processing usually leads to a loss of tocopherols (Wanasundra *et al.* 1998; Boki *et al.* 1991). However, in cases where chlorophyll compounds are abundant it is necessary to use activated earth to reduce the content.

To remove environmental pollutants active carbon has been used extensively in both vegetable and marine oil processing. Activated carbon used in bleaching is effective in decreasing dioxin and can remove as much as 90%, but is less efficient in removal of PCBs (Kock 2004; Maes *et al.* 2005). Unfortunately, activated carbon also very efficiently removes vitamin A that can act as antioxidant in natural oils.

4.2.4 Deodorisation

The deodorisation step removes undesirable ingredients from the oil and compounds formed during the preceding steps in the refining process. The deodorisation process is basically a steam distillation process, which will remove compounds that are more volatile than the triglycerides. The deodorisation of fish oil is often carried out at a temperature around 180 °C or above. Due to the high temperature peroxides are decomposed into secondary volatile oxidation products, which are then distilled (Bimbo 2007).

Batch or continuous deodorisers without efficient oil/surface ratio under processing will only remove parts of the contaminants. Batch deodorisation of fish oil at 180 °C for 2 hours showed only a 20% reduction in PCBs (Hilbert *et al.* 1998). To efficiently remove dioxin-like PCBs, distillation under high vacuum is necessary and can be done by either molecular distillation, thin film distillation or packed column stripping. All these process applications involve high temperatures between 150 and 230 °C and vacuum below 3 mbar. At temperatures over 180 °C the risk of isomerisation of cis- to trans fatty acids increases (Fournier *et al.* 2006). In patent application WO 2004007654 (Breivik and Thorstad 2004) an entrainer is used to facilitate the distillation of contaminants by adding fatty acid ethyl ester or other low molecular oil soluble components. This allows a reduced processing temperature under molecular distillation.

Improved designs of continuous deodorisation and structured packed columns have become more popular due to more efficient distillation and possibilities to reduce and change holding time at elevated temperatures. Although high efficiency is obtained there is an issue regarding the accumulation of decomposition products in the structured packing. Surface tension on wetted surfaces can hold several kilos of oil on its packed surface. Mechanical cleaning is almost impossible. During shutdowns small amounts of oxygen build up and will finally polymerise oil at the surface. Polymerised oil film will contaminate fresh oil in processing with taste and odour. The high temperature heating system has also been improved for more gentle heating of oil up to process temperature (Sjoberg *et al.* 1995). However, the main objective of deodorisation is still to remove breakdown peroxides and remove odour and taste to produce stable marine oil. Different methods for stabilising marine oils have been described with different approaches using a combination of processing aids and the addition of antioxidants during and after processing (Kendrick and Macfarlane 2001).

5. Optimising oxidative stability of bulk oil and omega-3 enriched products

The oxidative stability of bulk oils is influenced by several factors. However, when the oil is incorporated into foods, the complexity of the food matrix increases and so does the number of factors that may influence oxidative stability. Many foods are oil-in-water emulsions where the oil is emulsified into a water phase. In these foods, the composition of the oil-water interface is of particular importance as will be discussed later.

Table 1 summarises the most important factors that may affect the oxidative stability of bulk marine oils and omega-3 enriched food emulsions.

5.1 Oil quality

An important parameter to consider when developing omega-3 enriched foods is the quality, i.e. oxidative status of the omega-3 PUFA oil used for production as the oil quality may significantly influence the oxidative stability of the final omega-3 enriched product. Thus, it is important to consider all the factors important for the quality of the oil and the enriched product.

In fish oil enriched milk, addition of cod liver oil with a slightly elevated PV of 1.5 meq/kg oxidised significantly faster than a similar emulsion containing oil with a low PV of 0.1 meq/kg. Moreover, a sensory panel was able to distinguish milk emulsions produced with fish oil with a PV of 0.1 meq/kg as being less fishy and rancid as compared to milk produced with fish oil having a PV of 0.5 meq/kg

Table 1. *Important parameters influencing lipid oxidation in bulk fish oils for foods and food products enriched with fish oil.*

Parameter	Effect on oxidation in oil	Effect on oxidation in foods
Fish oil quality	Raw fish oil must be of high quality (i.e. low PV and AV) to obtain a good fish oil of eating quality	Generally PV should be below 1 meq/kg, for some foods an even lower PV may be required
Metal ions	Contact with trace metals should be avoided during processing of oils to reduce oxidation	Trace metals in food ingredients will in most cases increase oxidation and should therefore be reduced to a minimum
Oxygen	Reduce contact with air as much as possible	Reduce contact with air as much as possible
Temperature	Store oil at as low temperature as possible, preferably below 0 °C	Generally, temperature should be kept as low as possible. During processing of certain products higher temperature may result in better oxidative stability of the final product
Other lipid compounds	Free fatty acids and phospholipids should be removed during processing of oils	Free fatty acids will generally increase oxidation. Phospholipids may in certain cases act as antioxidants
Processing conditions	Important steps are degumming, neutralisation, bleaching and deodorisation.	Processing conditions of foods enriched with fish oil may need optimisation to reduce lipid oxidation
Endogenous antioxidants	A high content of endogenous antioxidants is preferable	A high content of endogenous antioxidants is generally preferable
Emulsifiers and composition of o/w interface	Not applicable	Choice of emulsifier should be considered when developing fish oil enriched foods based on an original recipe without fish oil
Packaging	Avoid light and air	Avoid light and air if possible

already after 1 day of storage (Let *et al.* 2005a). Yoghurt-based products enriched with omega-3 PUFA seems to be less susceptible to oxidation than omega-3 PUFA enriched milk (Let *et al.* 2007a). Thus, the importance of the oxidative quality of the oil used for production depends on the product in which it is used.

5.2 Metal ions

Iron has been shown to play an important role in the production of lipid radicals necessary to start and maintain the autooxidation process that leads to production of lipid peroxides. Other transition metal ions such as copper may also be involved in the metal catalysed decomposition of peroxides that propagates lipid oxidation as a chain reaction. Most oils contain metal ions (usually a Cu concentration in the µg/kg level and a Fe concentration in the mg/kg level) and trace levels will remain even after refining and deodorisation. In addition, in food emulsions many of the other ingredients may also contain trace levels of metal ions. Thus, since metals are so important for the oxidation reaction, it is important to minimise the contact between the metal ions and the oil. This contact can be minimised by complexing the iron by using chelating compounds such as EDTA or lactoferrin. Lactoferrin has been reported to suppress prooxidant activity of iron in fish oil and soybean oil in the temperature range 50-120 °C (Shiota *et al.* 2006). The use of lactoferrin and EDTA is however not permitted in all food products.

5.3 Oxygen

Oxygen is also vital for the oxidation reaction. The solubility of oxygen in oil is approx 5 times higher than in water. Thus, air should be excluded as much as possible during production and storage of oil and oil enriched products. However, in many cases, handling of oil and production of an o/w emulsion will result in the oil being exposed to air. It has been reported that a reduction of the access of oxygen to oil during production and storage retards lipid oxidation (Genot *et al.* 2003). Reduced exposure of the oil to oxygen can be achieved by processing under vacuum or in a nitrogen atmosphere. This would also reduce the amount of dissolved or trapped oxygen in the final product, which would otherwise promote oxidation. During storage, exclusion of headspace oxygen can be obtained by packaging in an air-tight container impermeable to oxygen under modified atmosphere or in vacuum.

5.4 Temperature

Heat is involved in the oxidation process of unsaturated fatty acids, thus, both processing and storage temperature should generally be as low as possible. The use of increased temperature is sometimes necessary during purification of omega-3 PUFA or during incorporation into emulsions, or is a result of mechanical processing during production. Thus, generally, this exposure to high temperature results in increased susceptibility of the lipids to oxidation. However, under some circumstances this is not the case. It has been observed that fish oil enriched milk oxidised faster, when emulsification of oil into milk was performed by homogenisation at high temperature compared to homogenisation at lower temperature (50 versus 72 °C) (Let *et al.* 2007b). This was explained by the unfolding of proteins at the higher temperature and thereby differences in protein composition of the oil-water interface. For autooxidation the oxidation rate doubles for each 10 degree increase in temperature. For light-induced oxidation the rate is not greatly influenced by temperature (Mozyraityte *et al.* 2006; Rahmani and Csallany 1998).

5.5 Other lipid compounds

Free fatty acids and mono- and diacylglycerols are breakdown products from triacylglycerol. They may act as prooxidants in oil by lowering the surface tension and thereby increasing the oxygen diffusion rate from headspace and into the bulk oil (Mistry and Min 1988). Therefore, their formation should be prevented or they should be removed as thoroughly as possible during oil refining to obtain stable oil.

As previously mentioned, phospholipids are generally removed during oil refining (degumming). They are reported to act both as pro- and antioxidants depending on their concentration and the presence of metal ions. Phospholipids have the ability to complex metal ions and may thereby inactivate prooxidative metal ions in the oil. On the other hand, the addition of phospholipids containing iron impurities, or not removing iron binding phospholipids during oil refining, will lead to a mistaken effect of phospholipids as prooxidants. As phospholipids also reduce the surface tension of edible oils, they will also increase the diffusion rate of oxygen from the headspace and into the oil, giving a prooxidative effect.

5.6 Processing conditions of food production

As mentioned previously, many of the same factors are important during the processing of oil or a product and during storage, i.e. avoiding exposure to air/oxygen, increased temperatures and contamination with metal ions. However, during processing these factors become increasingly important since many processes expose the oil to more vigorous conditions than during storage and it may be more difficult, for example, to minimise exposure to oxygen during processing compared to during storage. It is also important to consider that the oil/product may be contaminated with metal ions from the equipment used for processing.

Compared to vegetable oil, marine oils have low levels of endogenous antioxidants. The content may be even lower after purification of the oil. Thus, the endogenous content may not be sufficient to protect the oil against oxidation. Therefore, antioxidant addition may improve the oxidative stability of both bulk fish oil and omega-3 PUFA enriched as will be further discussed later.

5.7 Endogenous antioxidants

Antioxidants may often be present in a certain concentration range in order to work. The added antioxidants or ingredients may act synergistically with the endogenous antioxidants. Some antioxidants are able to regenerate other antioxidants (e.g. ascorbic acid may regenerate tocopherol). However, it is also important to consider the prooxidant effects of additive concentrations of endogenous and added antioxidants, since high concentrations of a given antioxidants may result in prooxidant effects. Thus, it is important to consider the concentration of both the endogenous antioxidant and the added antioxidant. The most important lipophilic antioxidants found in fish are tocopherol (primarily α-tocopherol) and carotenoids (primarily astaxanthin in coloured fish).

5.8 Emulsifiers and composition of oil-water interface

As mentioned before, the majority of the omega-3 enriched foods on the market today are emulsified products and in non-liquid food items emulsions may be used as a delivery system for omega-3 PUFA. Emulsions can either be water-in-oil (w/o) emulsions or, as for most of them, oil-in-water (o/w) emulsions. The physical structure of o/w emulsions is shown in Figure 3. Emulsions described hereafter will therefore be o/w emulsions. The emulsions consist of different phases: the continuous water phase, the oil phase and an interface between the oil droplets and the water phase.

Oil-in-water emulsions require the presence of emulsifiers in order to be physically stable. Emulsifiers are surface active molecules with amphiphilic properties, which make them capable of interacting with the oil-water interface whereby the surface tension is reduced. Examples of emulsifiers are macromolecules, such as proteins unfolding at the interface and smaller surfactant molecules, such as phospholipids, diacylglycerols, monoacylglycerols and free fatty acids.

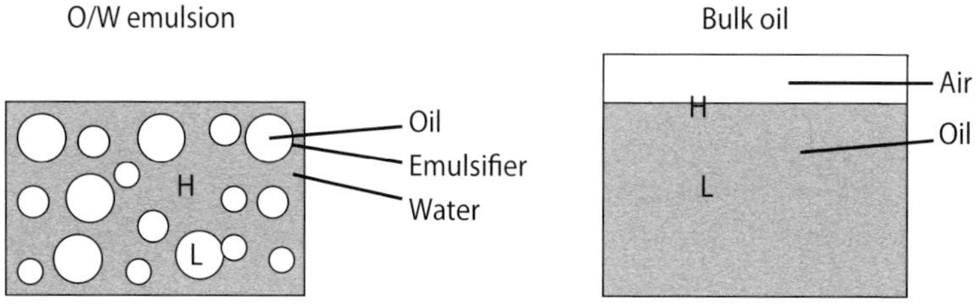

Figure 3. *Physical structure of o/w emulsion and location of a lipophilic (L) and hydrophilic antioxidant (H) in o/w emulsion versus bulk oil.*

Due to their location at the oil-water interface, emulsifiers have the ability to influence lipid oxidation in different ways by interacting with other ingredients as metal ions which are present in practically all food products. In protein stabilised emulsions, pH will generally be adjusted to avoid the coalescence of droplets. The resulting droplets will have a negatively charged surface that attracts metal ions or a positively charged surface that repels metal ions resulting in increased/decreased oxidation. Other surfactants such as phospholipids may also lead to charged oil droplets, thereby affecting oxidation. In addition to affecting the charge of emulsified droplets, the pH also affects the solubility of trace metals, which increases at decreasing pH (Belitz and Grosch 1999) and thereby potentially promotes oxidation. Therefore, careful selection of emulsifier and pH for production of omega-3 PUFA enriched foods is very important.

Not only the charge of the oil-water interface, but also its thickness seems to affect oxidation stability of emulsions. Thus, it is possible to create oil droplets coated with multiple layers of emulsifiers and thereby have a thicker and more resistant interface (Klinkesorn *et al.* 2005).

5.9 Packaging

Many products are packaged in transparent material, allowing the consumer to see the actual product. However, exposure of omega-3 PUFA to light will result in photo-oxidation (photosensitised oxidation). Light of a shorter wavelength has a more detrimental effect on oils than longer wavelengths (Sattar *et al.* 1976). The incorporation of UV-absorbing compounds in oil, like Tinuvin 234 or Tinuvin 236, packed in transparent plastic bottles improved sensory and oxidative stability of the oil (Azeredo *et al.* 2003). Thus, it is important to protect the omega-3 PUFA enriched foods from light during production and especially during storage.

6. Effect of antioxidants in bulk oil

6.1 Effect of antioxidants (synthetic and natural)

Among the various methods to prevent lipid oxidation, the addition of antioxidants is often considered and applied for stabilisation of marine oils intended for consumption. Recently, the interest in natural antioxidants has increased due to concerns about the long-term safety and negative perception of synthetic antioxidants. However, the most commonly used antioxidants are still synthetic because of their chemical stability, low cost and availability (Yang *et al.* 2002). Among the synthetic antioxidants butylated hydroxytoluene (BHT), butylated hydroxyanisole (BHA) and gallates are frequently used in the stabilisation of oils for consumption. BHT and BHA have shown synergistic effect and are available as commercial mixtures together with citric acid. Tertiary butyl hydroquinone (TBHQ), which is reported as an effective antioxidant, was until recently not approved in Europe, but as of July 2006 it is permissible to use this compound in a range of products.

Tocopherols are commercially available both as synthetic and natural compounds and several studies have been carried out to study the effect of different tocopherol (α-, β-, γ-, and δ-) homologues in fish oils (Hamilton *et al.* 1998; Han *et al.* 1997; Wada and Fang 1994; Zuta *et al.* 2007). α-tocopherol is found naturally in marine fish oils and this homologue is widely used as an antioxidant. The antioxidative effect of tocopherols in fish oils is concentration dependent (Kulås and Ackman 2001) and high concentrations (usually above 200 ppm) have shown no or negative effect in different fish oil systems (Falch *et al.* 2006; Kulås *et al.* 2002; Olsen *et al.* 2005; O'Sulluvan *et al.* 2005; Zuta *et al.* 2007). In bulk oil systems, several studies have shown that α-tocopherol may be less efficient than the other tocopherol homologues. Tocopherols have shown synergistic effects with other antioxidants (Kamal-Eldin and Appelqvist 1996) and different mixtures are commercially available such as mixtures with tocopherol, ascorbylpalmitate and lecithin. Olsen *et al.* (2005) demonstrated the synergistic effect of α-tocopherol used in combination with ascorbyl palmitate for the stabilisation of cod liver oil.

Recently also natural plant extracts such as rosemary, oregano, sage, flavonoids (Tsimidou *et al.* 1995) and green tea extracts (Frankel *et al.* 1996; Shahidi *et al.* 1997; Wanasundara and Shahidi 1994, 1998) were investigated for their effect in the stabilisation of fish oils. Rosemary has shown particularly good stabilising effect, better than other natural antioxidants in fish oils (Falch *et al.* 2006; O'Sullivan *et al.* 2005) and comparable with BHA and BHT. Rosemary is still the only herb commercially available for use as an antioxidant in Europe and the United States (Yanishlieva *et al.* 2006).

6.2 Stability of the commonly used antioxidants during processing

Thermal breakdown of tocopherols are reported to be relevant for temperatures above 260 °C (De Greyt *et al.* 1999) and modest losses are registered during the deodorisation/ distillation phase (carried out at 220-260 °C) (Kellens and De Greyt 2000; Rossi *et al.* 2001). Other studies have, however, shown that among the oil refining steps deodorisation is the refining step that leads to the highest loss of tocopherols, but that this also depends on the deodorisation method. The natural antioxidant activity of tocopherol should protect the refined oil against thermal oxidation.

6.3 Efficient analytical methods

There are several methods in use to evaluate the effect of antioxidants in bulk oils ranging from storage trials analysing the oil by traditional methods such as peroxide value, anisidine value, TBARS, headspace gas chromatographic techniques to accelerated methods. Accelerated methods may be: (1) the Schaal oven test (storage trial at 60 °C) followed by analysis of oxidation products; (2) active oxygen methods (measuring the drop in oxygen pressure); (3) conductivity measurements (OSI and Rancimat); and (4) free radical scavenging methods. Sensory analysis is probably the most valuable method, however it is expensive to use on a regular basis.

7. Effect of antioxidants in omega-3 PUFA enriched products

7.1 Emulsions versus bulk oil

The basic oxidation reactions of lipids present in emulsions are exactly the same as those of lipids in bulk oils. However, due to structural differences between emulsions and bulk oil systems the factors affecting lipid oxidation in these systems differ significantly. In emulsions, lipid oxidation is believed to be initiated at the interface between oil and water, but all phases of the system are involved, making the mechanisms of lipid oxidation in emulsions very different and more complex than those in bulk oil systems. In addition, several other intrinsic factors may affect lipid oxidation in emulsions, e.g. emulsion pH and transition metal ions present in ingredients in the aqueous phase and at the oil-water interface. Thus, special precautions have to be taken to avoid oxidative flavour deterioration of omega-3 PUFA enriched products. Also due to the structural differences antioxidant mechanisms in food emulsions are more complex than in bulk oils.

7.2 Antioxidant partitioning

The efficacy of antioxidants in emulsions depends on many factors, but partitioning of the antioxidant into the different phases seems to be one of the most important factors because the partitioning will determine the localisation of the antioxidants in multiphase systems, i.e. in the oil, the water or at the interface.

The relationship between antioxidant partitioning and antioxidant efficacy is also termed 'the polar paradox' as proposed by (Frankel *et al.* 1994; Huang *et al.* 1996; Porter 1993). According to the polar paradox, polar antioxidants are more active in non-polar media like bulk oils than their more non-polar counterparts, e.g. ascorbic acid and Trolox versus ascorbyl palmitate and tocopherol and vice versa. In emulsions, these effects have been suggested to be a result of the non-polar, lipid soluble antioxidants being located in the oil phase of the emulsion where oxidation propagates, whereas the polar antioxidants will be diluted in the continuous aqueous phase where their concentration will generally be too low to have any effect. The location of hydrophilic and lipophilic antioxidants in o/w emulsions vs. bulk oil is illustrated in Figure 3.

7.3 Other factors affecting antioxidant efficacy

Antioxidant efficacy is also affected by factors other than antioxidant partitioning. Thus, in different types of food emulsions the efficacy of antioxidants varies due to the different composition of the emulsions and the existence of different oxidation mechanisms (e.g. chain breaking, oxygen scavenging, metal chelation). Different effects of the same antioxidant are observed in different foods. For example, in fish oil enriched mayonnaise, tocopherol was inactive as an antioxidant and even seemed to have prooxidative effects at higher concentrations (>140 mg/kg oil) (Jacobsen *et al.* 2000, 2001), whereas, in fish oil enriched milk, γ-tocopherol (330 mg/kg), but not α-tocopherol was able to reduce lipid oxidation to some extent (Let *et al.* 2005b). Also ascorbic acid and ascorbyl palmitate were strong prooxidants in fish oil enriched mayonnaise. In contrast, ascorbyl palmitate was able to inhibit lipid oxidation to a much greater extent than γ-tocopherol in fish oil enriched milk (Let *et al.* 2005b). EDTA was demonstrated to be a prooxidant in model emulsions of fish or algae oil in water, when present in molar ratios of EDTA to iron of 1:1 or lower, but strongly inhibited oxidation at molar ratios of 2:1 and 4:1 (Frankel *et al.* 2002). On the other hand, in fish oil enriched mayonnaise a significant antioxidant effect of EDTA was observed at an EDTA:iron ratio of 1:2 (Nielsen *et al.* 2004).

In summary, it is not possible to use results obtained with one antioxidant in one product to predict the effect of the same antioxidant in another product.

8. Examples on applications

Omega-3 PUFA are available on the market in different forms such as ethyl esters, triacylglycerols and phospholipids. Ethyl esters are often concentrates of omega-3 PUFA used as dietary supplements. Phospholipids rich in omega-3 PUFA are also currently used for dietary supplements, but future applications in food products may also be a possibility. Omega-3 triacylglycerols are used for both dietary supplements and functional foods.

Omega-3 oils may be applied directly in the food as pure oils. However, in order to increase the oxidative stability of the oils they may also be incorporated into emulsions, microemulsions or microencapsulated powders with optimised properties that are subsequently used as delivery systems for omega-3 PUFA for foods. Microencapsulated oils are often used in baked products such as bread or in powdered products such as infant formula.

Omega-3 PUFAs have already been incorporated into a wide range of products that are commercially available. These products include dairy products such as milk and yoghurt, fitness bars, juice, infant formula and bread. Examples of product applications can be found at www.martek.com and www.denomega.com.

9. Conclusions

In conclusion, it is possible to produce high quality omega-3 PUFA oils that can be used for functional food products with good sensory properties and an acceptable shelf-life, but the control of raw material quality and lipid oxidation in oils and omega-3 enriched food products as well as contamination of the oils by pollutants will still pose a major challenge for the food industry in the years to come.

On the basis of the above review the following research and development needs can be identified:

Refining of oil

- The main need is to further improve removal of pollutants including lipophilic pollutants other than PCB and dioxins.
- For environmental reasons there is a trend towards a shift from chemical to physical refining, but further improvements in the physical refining method are necessary to optimise the quality of the refined oil.
- Methods to reduce PAH and formation of trans fatty acids during deodorisation should be developed.

Methods for determination and prediction of oxidation

- The methods available for determination of peroxides are either not specific or are time consuming. Therefore, fast and more specific methods for peroxides should be developed.
- Development of faster and at the same time more sensitive methods for volatiles. The current trend is that Solid Phase Microextraction (SPME) is increasingly being used, but this method is not as sensitive as the dynamic headspace method, so there is a need to further improve the sensitivity of this method.
- Lipid oxidation may be influenced by protein oxidation and vice versa, but sensitive and specific methods to determine protein oxidation are lacking and should be developed.
- Kinetic models for prediction of oxidation in complex systems should be developed in order to better predict the oxidation rate depending on important factors such as oxygen, temperature, pH, etc.

Oxidation and antioxidation mechanisms

- In complex food systems, lipid oxidation and antioxidant efficacy are influenced by a wide range of different factors and there is still a lack of understanding about which factors are the most important in different food systems.
- Research is needed to obtain a better understanding of the oxidation mechanisms in each different food system in order to be able to control oxidation more efficiently and to be able to extrapolate results obtained in one food product to other similar products.
- Research is also needed to improve our understanding of the mechanism of different antioxidants in different food products in order to be able to predict which antioxidants will work in which system and in which concentration.

References

Aaen B, Jessen F, Jensen B. 1995. Partial purification and characterization of a cellular acidic phospholipase A2 from cod (*Gadus morhua*) muscle. Comparative Biochemistry and Physiology Part B: Biochemistry and Molecular Biology 110:547-554.

Aubourg SP, Medina I. 1999. Influence of storage time and temperature on lipid deterioration during cod (*Gadus morhua*) and haddock (*Melanogrammus aeglefinus*) frozen storage. Journal of the Science of Food and Agriculture 79(13):1943-1948.

Aubourg SP, Sotelo CG. Perez-Martin R. 1998. Assessment of quality changes in frozen sardine (*Sardina pilchardus*) by fluorescence detection. Journal of the American Oil Chemists' Society 75(5):575-580.

Aubourg SP. 2001. Review: Loss of quality during the manufacture of canned fish products. Food Science and Technology International, 7(3):199-215.

Audley MA, Shetty KJ, Kinsella JE. 1978. Isolation and Properties of Phospholipase A from Pollock Muscle. Journal of Food Science 43:1771-1775.

Azeredo HMC, Faria JAF, Silva MAAP. 2003. The efficiency of TBHQ, β-carotene, citric acid and Tinuvin 234 on the sensory stability of soybean oil packaged in PET bottles. Journal of Food Science 68:302-306.

Baron CP, Andersen HJ. 2002. Myoglobin-induced lipid oxidation. A review. Journal of Agricultural and Food Chemistry, 50(14):3887-3897.

Barrier P. Rousseau J-Y. 1998. Method and plant for extracting fish oil and resulting products. WO 98/26674.

Belitz H-D, Grosch W. 1999. Food Chemistry. Berlin Heidelberg Germany: Springer Verlag.

Bimbo AP. 1990. Processing of fish oils'. In: Stansby ME (ed.) Fish oils in nutrition. New York, Van Nostrand Reinhold, p 181-225.

Bimbo AP. 2007. Processing of marine oils'. In: Breivik H (ed.) Long-Chaim Omega-3 Speciality Oils. Bridgwater England: The Oily Press, p. 77-110.

Boki K, Wada T, Ohno O. 1991. Effects of filtration through activated carbons on peroxides, thiobarbituric acid and carbonyl values of autoxidized soybean oil. JAOCS 68:561-565.

Breivik H, Thorstad O. 2004. A process for decreasing environmental pollutants in an oil or a fat, A volatile environmental pollutants decreasing working fluid, a health supplement, and an animal feed product. WO Patent 2007007654.

Burton GW, Ingold KU. 1984. beta-Carotene: an unusual type of lipid antioxidant. Science 224:569-573.

Chiu DT, van den Berg J, Kuypers FA, Hung IJ, Wei JS, Liu TZ. 1996. Correlation of membrane lipid peroxidation with oxidation of haemoglobin variants: possibly related to the rates of hemin release. Free Radical Biol Med 21:89-95.

Davies MJ. 2005. The oxidative environment and protein damage. Biochim Biophys Acta 1703(2):93-109.

De Greyt WF, Kellens MJ, Huynheraert AD. 1999. Effect of physical refining on selected minor components in vegetable oils. Fett/Lipid 101:428-432.

Decker EA, Erickson MC, Hultin HO. 1988. Enzymic lipid oxidative activities of sarcoplasmic reticulum in several species of northwest atlantic fish. Comparative Biochemistry and Physiology Part B: Biochemistry and Molecular Biology 91:7-9.

Elvevoll E. 1988. Biokjemikalier fra marint fett. Karakteriseringav fett fra torskeslo (*Gadus morhua*). Oppkonsentrering av ubiquinon med superkritisk fluid ekstraksjon (SFE). Dr. ing avhandling NTH, Trondheim.

European Commission, 2006, No 199/2006 of 3 February 2006 amending Regulation (EC) No 466/2001 setting maximum levels for certain contaminants in foodstuffs as regards dioxins and dioxin-like PCBs.

Falch E, Øverby A, Rustad T. 2006, Natural antioxidants in cod liver oil: pitfalls during oxidative stability assessment. In: Luten JB, Jacobsen C, Bekaert K, Sæbø A, Oehlschlager J (eds.) Seafood from fish to dish, Quality, safety and processing of wild and farmed fish. Wageningen, Wageningen Academic Publishers, p 127-136.

FAO 2004. The State of World Fisheries and Aquaculture. Rome FAO.

Fournier V, Destaillats F, Juaneda P, Dionisi F, Lambelet P, Sebedio JL, Berdeaux O. 2006. Thermal degradation of long-chain polyunsaturated fatty acids during deodorization of fish oil. European Journal of Lipid Science and Technology 108:33-42.

Frankel EN, Huang S-W, Aeschbach R, Prior E. 1996. Antioxidant activity of rosemary extract and its constituents, carsonic acid, carnosol and rosemarineic acid, in bulk and oil-in-water emulsion. Journal of Agricultural and Food Chemistry 44:131-135.

Frankel EN, Huang S-W, Kanner J, German JB. 1994. Interfacial phenomena in the evaluation of antioxidants: Bulk oils vs emulsions. Journal of Agricultural and Food Chemistry 42:1054-1059.

Frankel EN, Satue-Gracia T, Meyer AS, German JB. 2002. Oxidative stability of fish and algae oils containing long-chain polyunsaturated fatty acids in bulk and in oil-in-water emulsions. Journal of Agricultural and Food Chemistry 50:2094-2099.

Funes JA, Weiss U, Karel M. 1982. Effects of reaction conditions and reactant concentrations on polymerization of lysozyme reacted with peroxidizing lipids. Journal of Agricultural and Food Chemistry. 1082(30):1204-1208.

Gbogouri GA, Linder M, Fanni J, Parmentier M. 2006. Analysis of lipids extracted from salmon (Salmo salar) heads by commercial proteolytic enzymes. European Journal of Lipid Science and Technology 108:766-775.

Genot C, Meynier A, Riaublanc A. 2003. Lipid oxidation in emulsions. In: Kamal-Eldin A (ed.) Lipid oxidation pathways. AOCS Press: Champaign, IL p. 190-244.

GRP. NN. 2007. World Surimi Market. GRP 89 www.globefish.org

Gunstone ED, Harwood JL, Padley FB. 1994. The Lipid Handbook, Cambridge, Cambridge University Press p. 167-179.

Hamilton RJ, Kalu C, McNeill GP, Padley FB, Pierce JH, 1998. Effects of tocopherols, ascorbyl palmitate and lecithin on the autoxidation of fish oil. Journal of the American Oil Chemists' Society 75(7):813-822.

Han D, Shin H-K, Yoon SH. 1997. Separation of w-3 polyunsaturated fatty acids from fish oil and stabilisation of the oil against oxidation. In: Shahidi F, Cadwallader KR (ed.) Flavour and Lipid Chemistry of Seafood. ACS Symposium series 674, Washington DC, p. 255-263.

Hernandez E, Rathbone SJ. 2002. Refining of glyceride oils by treatment with silicate solutions and filtration. US Patent 6 448 423,

Hilbert G, Lillemark L, Balchen S, Højskov CS. 1998. Reduction of organochlorine contaminants from fish oil during refining. Journal of Chemosphere 37(7):1241-1252.

Hruschka S, Frische R. 1998. A new oil extraction process: FRIOLEX (R). Ocl-Oleagineux Corps Gras Lipides 5:356-360.

Huang S-W, Hopia A, Schwarz K, Frankel EN, German JB. 1996. Antioxidant activity of á-tocopherol and trolox in different lipid substrates: Bulk oils vs oil-in-water emulsions. Journal of Agricultural and Food Chemistry 44:444-452.

Ingemansson T, Olsson NU, Herslof BG, Ekstrand B. 1991. Lipids in Light and Dark Muscle of Farmed Rainbow-Trout (*Oncorhynchus-Mykiss*). Journal of the Science of Food and Agriculture 57:443-447.

Ingemansson T, Pettersson A, Kaufmann P. 1993. Lipid Hydrolysis and Oxidation Related to Astaxanthin Content in Light and Dark Muscle of Frozen Stored Rainbow-Trout (*Oncorhynchus-Mykiss*). Journal of Food Science 58:513-518.

Jacobsen C, Hartvigsen K, Lund P, Adler-Nissen J, Holmer G, Meyer AS. 2000. Oxidation in fish-oil-enriched mayonnaise 2. Assessment of the efficacy of different tocopherol antioxidant systems by discriminant partial least squares regression analysis. European Food Research and Technology 210:242-257.

Jacobsen C, Hartvigsen K, Lund P, Thomsen MK, Skibsted LH, Holmer G, Adler-Nissen J, Meyer AS. 2001. Oxidation in fish oil-enriched mayonnaise: 4. Effect of tocopherol concentration on oxidative deterioration. European Food Research and Technology 212:308-318.

Jansson S, Elveveoll E. 2000. Process for separating lipids and proteins from biological material.WO/2000/023545, http://www.wipo.int/pctdb/en/wo.jsp?wo=2000023545

Kamal-Eldin A, Appelqvist L. 1996. The chemistry and antioxidant properties of tocopherols and tocotrienols. Lipids 31:671-701.

Kellens M, De Greyt W. 2000. Deodorization. In: O'Brien R, Farr W, Wan P (eds.) Introduction to fats and oils technology. AOCS Press, Champaign, IL. p 235-268.

Kendrick A, Macfarlane N. 2001. Preparation and stabilization of food-grade marine oils. US patent 6 623 774.

Klinkesorn U, Sophanodora P, Chinachoti P, McClements DJ, Decker EA. 2005. Increasing the oxidative stability of liquid and dried tuna oil-in-water emulsions with electrostatic layer-by-layer deposition technology. Journal of Agricultural and Food Chemistry 53:4561-4566.

Kock JD. 2004. Removal of Dioxins and PCBs from Marine Oils: Current Status and Future Developments. Aqua Feeds: Formulation and Beyond 1(2):9-10.

Kulås E, Ackman E. 2001. Protection of α-tocopherol in non purified and purified fish oil. Journal of the American Oil Chemists' Society 78:197-203.

Kulås E. Olsen E, Ackman E. 2002. Effect of alpha-, gamma-, and delta-tocopherol on the distribution of volatile secondary oxidation products in fish oil. Eur. J. Lipid Sci. Technol. 104:520-529.

Let MB, Jacobsen C, Meyer AS. 2005a. Sensory stability and oxidation of fish oil enriched milk is affected by milk storage temperature and oil quality. International Dairy Journal 15:173-182.

Let MB, Jacobsen C, Pham KA, Meyer AS. 2005b. Protection against oxidation of fish-oil-enriched milk emulsions through addition of rapeseed oil or antioxidants. Journal of Agricultural and Food Chemistry 53:5429-5437.

Let MB, Jacobsen C, Meyer AS. 2007a. Lipid oxidation in milk, yoghurt, and salad dressing enriched with neat fish oil or pre-emulsified fish oil. J. Agric. Food Chem. 55:7802-7809.

Let MB, Jacobsen C, Sørensen ADS, Meyer AS. 2007b. Homogenization conditions affects the oxidative stability of fish oil enriched milk emulsions: Lipid oxidation. J Agric Food Chem. 55:1773-1780.

Liu W, Wang JY. 2005. Modifications of protein by polyunsaturated fatty acid ester peroxidation products. Biochim Biophys Acta 1752(1):93-98.

Lopez-Amaya C, Marangoni AG. 2000. Lipases. In: Haard NF, Simpson BK (eds.) Seafood Enzymes. Utilization and Influence on Postharvest Quality. New York: Marcel Dekker Inc. p. 121-146.

Maes J, De Meulenaer B, Van Heerswynghels P, De Greyt W, Eppe G, De Pauw E, Huyghebaert A. 2005. Removal of dioxins and PCB from fish oil by activated carbon and its influence on the nutritional quality of the oil. Journal of the American Oil Chemists' Society 82:593-597.

Mistry BS, Min DB. 1988. Prooxidant effects of monoglycerides and diglycerides in soybean oil. J Food Sci 53:1986-1987.

Mozyraityte R, Rustad T, Storrø I. 2006. Pro-oxidant activity of Fe^{2+} in oxidation of cod phospholipids in liposomes. Eur. J. Lipid Sci. Technol. 108:218-226.

Neas NP, Hazel JR. 1984. Temperature-Dependent Deacylation of molecularspecies of phosphatidylcholine by microsomal Phospholipase A_2 of Thermally Acclimated Rainbow Trout *(Salmo gairdneri)*. Lipids 19:258-263.

Nielsen NS, Petersen A, Meyer A, Timm-Heinrich M, Jacobsen C. 2004. Effects of lactoferrin, phytic acid and EDTA on oxidation in two food emulsions enriched with long chain polyunsaturated fatty acids. Journal of Agricultural and Food Chemistry 52:7690-7699.

O`Sullivan A, Mayr A, Shaw NB, Murphy SC, Kerry JP. 2005. Use of Natural antioxidants to stabilize fish oil systems. Journal of Aquatic Food Product Technology 14(3):75-94.

Olsen E, Vogt G, Saarem K, Greibrokk T, Nilsson A. 2005. Autoxidation of cod liver oil with tocopherol and ascorbylpalmitate. JAOCS, 82(2):97-103.

Pan BS, Kuo JM. 2000. Lipoxygenase. In: Haard NF and Simpson BK (eds.) Seafood Enzymes. Utilization and Influence on Postharvest Quality. New York: Marcel Dekker Inc. p 317-336.

Passi S, Cataudella S, Tiano L, Littarru GP. 2005. Dynamics of lipid oxidation and antioxidant depletion in Mediterranean fish stored at different temperatures. Biofactors 25:241-254.

Porter WL. 1993. Paradoxical behavior of antioxidants in food and biological systems. Toxicol. Ind. Health 9:93-122.

Rahmani M, Csallany SA. 1998. Role of minor constituents in the photooxidation of virgin olive oil. Journal of the American Oil Chemists' Society 75:837-843.

Rampon V, Lethuaut L, Mouhous-Riou N, Genot, C. 2001. Interface characterization and aging of bovine serum albumin stabilized oil-in-water emulsions as revealed by front-surface fluorescence. Journal of Agricultural and Food Chemistry 49(8):4046-4051.

Recommendation 2004/705/EC monitoring background levels of dioxins and dioxin like PCBs in foodstuffs.

Richards MP, Hultin HO. 2002. Contributions of blood and blood components to lipid oxidation in fish muscle. Journal of Agricultural and Food Chemistry 50:555-564.

Rossi M, Gianazza M., Alamprese C, Stanga F. 2001. The effect of bleaching and physical refining on color and minor components of palm oil. JAOCS 78(10):1051-1055.

Sattar A, DeMan JM, Alexander JC. 1976. Effect of wavelength on light induced quality deterioration of edible oils and fats. Can Inst Food Sci. Technol. J. 9:108-113.

Shahidi F, Wanasundara UN, He Y, Shukla VKS. 1997. Marine Lipids and their Stabilization with Green Tea and Catechins. In: Shahidi F, Caldwallader K (eds.) Flavour and Lipid Chemistry of Seafoods. American Chemical Society Symposium series 674, Washington DC, p 186-197.

Sheridan MA. 1988. Lipid dynamics in fish aspects of absorption, transportation, deposition and mobilisation. Comparative Biochemistry and Physiology Part B 90:679-690.

Shiota M, Uchida T, Oda T, Kawakami H. 2006. Utilization of lactoferrin as an iron stabilizer for soybean and fish oil. J Food Science 71:C120-123.

Shirai N, Suzuki H, Tokairin S, Ehara H, Wada S. 2002. Dietary and seasonal effects on the dorsal meat lipid composition of Japanese *(Silurus asotus)* and Thai catfish (*Clarias macrocephalus* and hybrid *Clarias macrocephalus* and *Clarias galipinus*). Comparative Biochemistry and Physiology A-Molecular and Integrative Physiology 132:609-619.

Sikorski ZE, Kolakowski E. 2000. Endogenous Enzyme Activity and Seafood Quality: Influence of Chilling, Freezing, and Other Environmental Factors. In: Haard NF, Simpson BK (eds.) Seafood Enzymes. Utilization and Influence on Postharvest Quality. New York: Marcel Dekker Inc. p. 451-487.

Silkeberg, Kochhar A, Parkash S. 1998. Refining of edible oil retaining maximum antioxidative potency. US Patent 6 033 706.

Sjoberg, Hillstrom P, Anders MA. 1995. Vacuum vessel for continuous or semicontinuous treatment of fatty oils. US Patent 6 165 325.

Sovik SL, Rustad T. 2005. Effect of season and fishing ground on the activity of lipases in byproducts from cod (*Gadus morhua*). LWT - Food Science and Technology 38:867-876.

Toeneboehn GJ, Walter M, Cheek III, Welsh AW, Bogdanor JM. 1994. MPR process for treating glyceride oils, fatty chemicals and wax. US Patent 5 298 639.

Trenzado C, Hidalgo MC, Garcia-Gallego M, Morales AE, Furne M, Domezain A, Domezain J, Sanz A. 2006. Antioxidant enzymes and lipid peroxidation in sturgeon *Acipenser naccarii* and trout *Oncorhynchus mykiss*. A comparative study. Aquaculture 254:758-767.

Tsimidou M, Papavergou E, Boskou D. 1995. Evaluation of oregano antioxidant activity in mackerel oil. Food Research International 28(4):431-433.

Undeland I, Ekstrand B, Lingnert H. 1998. Lipid oxidation in herring (*Clupea harengus*) light muscle, dark muscle, and skin, stored separately or as intact fillets. Journal of the American Oil Chemists Society 75:581-590.

Undeland I, Richards MP, Hultin HO. 2002. Added triacylglycerols do not hasten hemoglobin-mediated lipid oxidation in washed minced cod muscle. J Agric Food Chem 50(23):6847-6853.

Undeland I., Hall G, Lingnert H. 1999. Lipid Oxidation in Fillets of Herring (*Clupea harengus*) during Ice Storage. J Agric Food Chem 47:524-532.

Wada S, Fang X. 1994. Synergistic antioxidant effect of rosemary and α- tocopherols at different storage temperatures and its applications for inhibiting dried sardine meat oxidation, Journal of the Japan Oil Chemists' Society 43:109-115.

Wanasundara UN, Shahidi F. 1994. Stabilisation of seal blubber and menhaden oil with green tea, JAOCS 73:1183-1190.

Wanasundara UN, Shahidi F, Amarowicz R. 1998. Effect of processing on constituents and oxidative stability of marine oils, Journal of Food and Nutrition 5:29-41.

Wanasundara UN, Shahidi F. 1998. Antioxidant and pro-oxidant of green tea extracts in marine oils, Food Chem. 63(3):335-342.

Welsh AW, Bogdanor JM, Toeneboehn GJ. 1989. Silica refining of oils and fats - World conference proceedings – Basic principles and modern practices. American Oil Chemist Society. P. 189-202.

Yang M-H, Lin H-J, Choong, Y-M. 2002. A rapid gas chromatographic method for direct determination of BHA, BHT and TBHQ in edible oils and fats. Food Research International 35(7):627-633.

Yanishlieva NV, Marinova E, Pokorny J. 2006. Natural antioxidants in herbs and spices. Eur. J. Lipid Sci. and Tech. 108:776-793.

Yates RA, Caldwell JD. 2000. Synthetic Magnesium Silicate Filter Media - Effects on Frying Oil Quality - Joint International Congress and Expo Lipids, Fats and Oils Opportunities and Responsibilities in the New Century Würzburg Germany. October 8-10. (Book of abstract. Poster V/15), www.eurofedlipid.org/meetings/archive/wuerzburg/wb2000_poster.pdf

Zuta CP, Simpson BK, Zhao X, Lercerc L. 2007. The effect of α-tocopherol on the oxidation of mackerel oil. Food Chemistry 100(2): 800-807.

Fish proteins and peptide products: processing methods, quality and functional properties

***Gudjon Thorkelsson*[1], *Rasa Slizyte*[2], *Asbjørn Gildberg*[3] *and Hordur G. Kristinsson*[1,4]**
[1]*Matis, University of Iceland, Skulagata 4, IS 101 Reykjavik, Iceland*
[2]*SINTEF Fisheries and Aquaculture, N-7465 Trondheim, Norway*
[3]*Nofima Marine, POBox 6122, N-9291 Tromsø, Norway*
[4]*University of Florida, Department of Food Science and Human Nutrition, Gainesville, Florida 32611, USA*

1. Introduction

Fish meal, fish sauce, surimi and fish silage are traditional protein-based products (Rustad 2003; FAO 2004; Dissaraphong *et al.* 2006; Vidal-Giraud and Chateau 2007; FIN 2007). The evidence that marine proteins, protein hydrolysates and peptides also provide a number of health beneficial effects is growing and therefore marine proteins also have great potential as ingredients in functional foods (Undeland *et al.* 2009). When applying marine proteins in foods it is of the utmost importance that they have a high quality so that the sensory and functional properties and stability of the final product are acceptable.

This chapter gives an overview of the methods used for processing marine proteins and peptides and of the factors that can affect the quality and physical properties of proteins used as ingredients in functional foods. It also gives information about seafood protein and peptide products on the market, and the future challenges in research and development in this area are discussed.

2. Sources of raw material

Selection, storage and handling of the raw material are preliminary steps in the processing of fish protein and peptide ingredients. Seafood-based raw materials are unstable and very perishable. Spoilage, development of off-flavours and odours must be prevented by controlling protein degradation and lipid oxidation. Raw material containing chemical contaminants like PCBs and dioxins exceeding maximum limits and action limits set by authorities must be avoided (Anonymous 2004). Furthermore, development of biogenic amines can limit the use of certain species and care must be taken to use only fresh unspoiled raw material.

The raw material can be underutilised fish and by-products like viscera, backbones, skins, cut-offs, saw dust, wash water and cooking juices. Collection of raw material to centralised production facilities is difficult due to spoilage and transportation cost.

The fish protein supply is very diverse compared to other protein sources. The diversity is an opportunity, but can also create problems and make things difficult. Besides species diversity, there is diversity in fishing vessels, production and processing sites and operations. This can create difficulties in collecting by-products and lead to variable quality and conditions of the raw materials, making them unfit for processing into high-value products. One of the main criteria for the production of fish proteins is that the underutilised species and by-products receive the same treatment as the main product, and that they are processed when they are as fresh as possible.

A stable and sustainable supply is necessary to start up businesses producing fish proteins and peptides. The conditions of the wild fish stocks and the seasonality of the catches must be considered and can be a problem when planning large-scale production of proteins and peptides. Many of the fish stocks are declining while others are in good condition and even increasing. In Iceland, for example, pelagic fish represents about 70% of the total catch. Capelin catches have been going down, while blue whiting and herring catches have been increasing. The catch is seasonal, with capelin being mostly caught during the first months of the year, blue whiting in early spring and summer and herring in the summer and autumn. Groundfish like cod, haddock and saithe are however landed throughout the year and there is far less fluctuation in the size of the stocks than for the pelagic fish species.

3. Microbiology of raw material for fish protein and peptide processing

3.1 Chilling, handling and spoilage

Rapid chilling, storage at low temperature, hygiene and a short time between catch and processing are crucial factors in controlling both the amount, growth and activities of spoilage bacteria. Fish spoil at very different rates. For example, cod spoils rapidly compared to flatfish. Each fish product has its own specific spoilage bacteria. *Psychrotrophic* bacteria belonging to *Pseudomonas spp.* and *Shewanella putrefaciens* dominate the spoilage flora of iced stored fish. *Shewanella* spoilage is characterised by TMA and sulphides (H_2S) whereas the *Pseudomonas* spoilage is characterized by the absence of these compounds and occurrence of sweet, rotten sulphydryl odours. The microflora changes with storage temperature. At high storage temperatures (15-30 °C) different species of Vibrionaceae, Enterobacteriaceae and Gram-positive organisms are responsible for spoilage. Thus, the temperature is very important and the shelf-life of fish products, which is markedly extended when products are stored at low temperatures (Huss 1995).

Gutting improves quality and the storage life of many fish species. During feeding periods the fish contain many bacteria producing digestive enzymes. In most cases, small- and medium-sized fatty fish such as herring, sardines and mackerel are not eviscerated immediately after catch. Problems may arise with ungutted fish during periods of heavy feeding due to belly-bursting. Autolysis 'self-digestion' can also be a problem. Physical handling of the fish accelerates this process. The low molecular weight peptides and free amino acids produced by the autolysis of proteins not only lower the commercial acceptability of pelagic fish, but in bulk-stored capelin, autolysis has been shown to accelerate the growth of spoilage bacteria. Rough handling will result in a faster spoilage rate, due to the physical damage to the fish, resulting in easy access for enzymes and spoilage bacteria.

Histamine formation due to microbial activity in fish belonging to the *Scombroidae* and *Scomberesocidae* family (tuna, mackerel, skipjack, bonito) as well as certain non-scombroid fish (herring, anchovies, sardines) can lead to poisoning if it is not controlled. Chilling seafood to 2-5 °C eliminates histamine formation by the mesophilic bacteria. However, psychrotolerant bacteria can produce toxic concentrations of histamine in seafood at 2-5 °C. In order to control histamine formation in chilled seafood, storage conditions as well as product characteristics and shelf-life must be carefully selected (Emborg *et al.* 2006).

3.2 Microbial criteria for raw material, processes and ready products

There are general food safety requirements, according to which food must not be placed on the market if it is unsafe. Food processing operators have an obligation to withdraw unsafe food from the market (Anonymous 2002). There are also rules that lay down microbiological criteria for foodstuffs (Anonymous 2005) including products with fish protein and fish protein hydrolysates. The rules contain harmonised safety criteria on the acceptability of food, in particular as regards the presence of certain pathogenic micro-organisms. Microbiological criteria also give guidance on the acceptability of foodstuffs and their manufacturing, handling and distribution processes. The use of microbiological criteria should form an integral part of the implementation of HACCP-based procedures and other hygiene control measures. Microbiological criteria can be used in validation and verification of HACCP procedures and other hygiene control measures. The microbiological criteria define the acceptability of the processes, and also food safety microbiological criteria set a limit above which a foodstuff should be considered unacceptably contaminated with the microorganisms for which the criteria are set. Criteria and action levels for pathogens and histamines in products that could contain fish proteins and fish protein hydrolysates are shown in Table 1. Hygienic and pathogen guidelines from the Icelandic Matis microbial laboratory for both raw material and protein powders are listed in Table 2.

Table 1. *Criteria for pathogens and histamines in products that could contain fish proteins and fish protein hydrolysates.*

Food category	Micro-organisms/their toxins, metabolites	Sampling plan		Limits		Stage where the criterion applies
		n	c	m	M	
Ready-to-eat foods intended for infants and ready-to-eat foods for special medical purposes	*Listeria monocytogenes*	10	0	Absence in 25 g		Products placed on the market during their shelf-life
Ready-to-eat foods able to support the growth of *L. monocytogenes*, other than those intended for infants and for special medical purposes	*Listeria monocytogenes*	5	0	100 cfu/g		Products placed on the market during their shelf-life
	Listeria monocytogenes	5	0	Absence in 25 g		Before the food has left the immediate control of the food business operator, who has produced it
Ready-to-eat foods unable to support the growth of *L. monocytogenes*, other than those intended for infants and for special medical purposes	*Listeria monocytogenes*	5	0	100 cfu/g		Products placed on the market during their shelf-life
Gelatine and collagen	*Salmonella*	5	0	Absence in 25 g		Products placed on the market during their shelf-life
Cooked crustaceans and molluscan shellfish	*Salmonella*	5	0	Absence in 25 g		Products placed on the market during their shelf-life
Dried infant formulae and dried dietary foods for special medical purposes intended for infants below six months of age	*Salmonella*	30	0	Absence in 25 g		Products placed on the market during their shelf-life
Dried infant formulae and dried dietary foods for special medical purposes intended for infants below six months of age	*Enterobacter sakazakii*	30	0	Absence in 10 g		Products placed on the market during their shelf-life
Fishery products from fish species associated with a high amount of histidine	Histamine	9	2	100 mg/kg	200 mg/kg	Products placed on the market during their shelf-life
Fishery products which have undergone enzyme maturation treatment in brine, manufactured from fish species associated with a high amount of histidine	Histamine	9	2	200 mg/kg	400 mg/kg	Products placed on the market during their shelf-life

n = number of units comprising the sample; c = number of sample units giving values between m and M.
Based upon Anonymous 2005.

Table 2. *Hygienic and pathogen guidelines from the Icelandic Matis microbial laboratory.*

Criteria	Micro-organisms/their toxins, metabolites	Sampling plan		Limits	
		n	c	m	M
Raw and fresh fish, fillets and cuts.	Total viable count 22 °C	5	2	$2.5x10^5$/g	$5x10^5$/g
	Total viable count 35 °C	5	2	$1x10^5$/g	$2x10^5$/g
	E. coli	5	2	$1x10^2$/g	$2x10^2$/g
	Faecal *E. coli*	5	2	0	4
	Staphylococcus aureus	5	2	$1x10^1$/g	$1x10^2$/g
Minced fish and homogenised fish	Total viable count 30 °C	5	2	$1x10^{6/g}$	$1x10^7$/g
	Total viable count 17 °C and 7 °C	5	2	$1x10^{6/g}$	$1x10^7$/g
	Faecal *E. coli*	5	2	$1x10^1$/g	$1x10^2$/g
Protein powders	*E. coli* 30 °C	5	2	0/g	$1x10^1$/g
	Staphylococcus Aureus	5	2	$1x10^1$/g	$1x10^2$/g
	Bacillus cereus	5	2	$1x10^2$/g	$1x10^3$/g
	Listeria monocytogenes	5	2	Not present	
	Salmonella	10	2	Not present	

4. Factors affecting quality of raw materials

Oxidation, enzymatic processes, heating, reactions with lipid oxidation products, changes in pH and freezing of the raw material can cause protein changes such as denaturation, formation of aggregates, cross linking and breakdown of polypeptide chains. These changes can lead to reduced nutritional value due to changes in amino acids or reduced digestibility. They can also lead to changes in colour, solubility and other functional properties. The rates of most of these changes are highly temperature dependent and chilling and freezing reduce the reaction rate. The best way to slow or prevent them is to use fresh raw material and process it as soon as possible in the most hygienic manner.

4.1 Protein oxidation and interactions with lipids

Marine proteins are vulnerable to oxidative attack during processing and storage. Reactive oxygen species (ROS) are mainly responsible for initiation of oxidation reactions of food (Choe and Min 2006). Oxidation can result in major physical changes in protein structure and may lead to reduced nutritional value due to the blocking of ε-amino groups of L-lysine or changes in other amino acids, especially tryptophan, cysteine, cystine, histidine, methionine and tyrosine. The most important ROS are hydroxyl radicals and singlet oxygen (Li and King 1999; Davies 2005). Singlet oxygen easily reacts with aromatic amino acids (Choe and Min 2006). Protein oxidation by hydroxyl radical produces cross-linked products and carbonyl compounds. Denaturation by ʻOH involves protein unfolding and increased accessibility of peptide bonds of proteases (Davies 1987).

It has been shown that lipid oxidation products can react with proteins and lead to Schiff base formation (Yong and Karel 1978), cross linking (Li and King 1999; Tironi *et al.* 2002), polypeptide chain scission (Li and King 1999), denaturation and destruction of some amino acids (Soyer and Hultin 2000; Hidalgo and Zamora 2002; Refsgaard *et al.* 2000). Mild oxidation can improve gelling properties or have no effect, while more oxidised lipid could reduce gel forming ability (Tunhun *et al.* 2002). Lipid oxidation may be one of the factors contributing to the loss of protein solubility during frozen storage (Saeed and Howell 2002). The secondary products arising from hydroperoxide decomposition also readily damage protein and amino acids through formation of covalent bonds. Lipid-protein interaction is dependent on oxygen levels, formation of lipid oxidation products (Aubourg and Medina 1999), storage temperature and time (Aubourg *et al.* 1998; Aubourg and Medina 1999; Rampon *et al.* 2001) degree of unsaturation of lipid, contact area between lipid and protein (Rampon *et al.* 2001) and concentration, amount of prooxidants and antioxidants.

Both metmyoglobin and oxymyoglobin have been shown to be major initiators of lipid oxidation and peroxidation. Oxidation/peroxidation induced by the presence of myoglobin is highly pH dependent (Baron and Andersen 2002). It has also been shown that proteins can inhibit or slow down lipid oxidation. Fish protein hydrolysates and the number of peptides obtained from fish proteins have shown antioxidative properties. Antioxidative mechanisms are probably related to the ability of peptides and hydrolysates to scavenge radicals formed during peroxidation and chelate transition metals, which are potent prooxidants (Guerard *et al.* 2007).

4.2 Other post mortem biochemical changes in fish

There are rapid changes in the properties/quality of fish after harvesting due to enzymatic processes caused by endogenous (autolytic process) and bacterial enzymes influencing texture, flavour and odour. Lysosomes contain numerous enzymes capable of degrading both carbohydrates, lipids, nucleic acids and proteins. Many types of proteolytic enzymes have been found in fish muscle and other fish tissues. Proteolytic breakdown activated by pH drop caused by glycolysis results in softening of fish muscle and in formation of peptides and free amino acids.

Endogenous enzymes (lipases, phospholipases; lipoxygenases, peroxidases) are still active during frozen storage, especially if light or other catalysts (heme groups, transition metals) are present. As a result, a wide range of oxidation and hydrolysis compounds are produced, most of them relatively small molecules, that are susceptible to interaction with fish constituents leading to a lowering in nutritional and sensory values of the product (Aubourg 2001).

One of the most notable examples of autolytic proteolysis is the incidence of belly-bursting in pelagic species. *Post mortem* degradation of fish muscle is probably the result of the action both of calpains and cathepsins working together to break down myofibrillar proteins while collagenases break down components in the extracellular matrix.

The reduction of trimethylamine oxide (TMAO) is usually due to bacterial action but in some species (mainly gadoid species) TMAOase or TMAO demethylase in the muscle tissue will break down TMAO to dimethylamine (DMA) and formaldehyde (FA). Formaldehyde may induce crosslinking of the muscle proteins making the muscle tough and reducing water holding capacity. Most of the TMAO demethylase enzymes are membrane-bound and become active when the tissue membranes are disrupted by freezing or artificially by detergent solubilisation. To prevent production of formaldehyde, fish should be stored at stable temperatures <-30 °C and rough handling or application of physical pressure on the fish prior to freezing should be avoided.

5. Fish protein processing methods

There are two main processing methods of fish proteins for human consumption, firstly, methods that improve yield and utilisation of proteins in traditional fish processing and, secondly, methods to produce marine proteins and peptides with functional and bioactive properties. It is important to include methods that improve yield for human consumption in this review as they are very important in increasing the supply of fish proteins for direct human consumption, and there is growing evidence on the specific health effects of fish proteins *per se* (Thorsdottir and Ramel 2008).

5.1 Surimi vs. pH shift

Fish protein isolates from trimmings and cut-offs can be used to increase the yield of raw material in fish filleting operation and in the production of ready-to-eat seafood products. This is done to improve quality, firmness and juiciness and to meet increasing demand for convenience food items while at the same time increasing yield (Xiong 2005). There are two technologies for improving fillet yield. The SuspenTec® system, acid and alkali extraction (pH-shift) are now in use in the fish industry (Batista 1999; Hultin and Kelleher 1999; Hultin *et al.* 2004; Hultin *et al.*

2005; Shahidi 2006; Kelleher and Williamson 2007). The SuspenTec® process is an automated method of reducing fish trimmings at low temperatures (-(4-6) °C) to small particles and incorporating them into traditional brines. The low temperature ensures efficient protein binding and dispersal of suspension into the whole-muscle product. The pH-shift protein isolate can be added to fresh seafood by needle injection into fillets, vacuum tumbling or emulsification. Acid and alkali extracted fish protein isolates have a GRAS (Generally Regarded as Safe) status in the United States (FDA 2004).

The alkali and acid pH-shift methods have been compared with traditional surimi processing. They are claimed to be more suitable than the surimi process for complex raw materials like whole fish and co-products. The processes include solubilising muscle proteins by treating diluted finely homogenised fish meat to either very low pH or a very high pH at low temperatures. The next step is centrifugation to remove insoluble solids. The soluble proteins are then precipitated by bringing the pH to the isoelectric point of the myofibrillar proteins to yield a protein isolate (Kristinsson *et al.* 2006).

The main difference between the methods is that sarcoplasmic proteins are washed away during surimi processing but recovered during pH-shift processing resulting in a higher yield (Kristinsson *et al.* 2006). The protein isolate from the pH-shift process is also claimed to have a substantial absence of both neutral and membrane lipids that is expected to greatly improve the oxidative stability of the product. The alkali-aided process gives a more oxidative stable isolate than the acid-aided process and is sometimes more stable than surimi. Haemoglobin, a main catalyst of oxidation in fish muscle, is highly prooxidative at low pH but it is stabilised at high pH (Kristinsson and Hultin 2004). Heme proteins are also more actively removed from the alkaline process than from surimi, yielding a whiter product which is more stable to oxidation (Kristinsson 2002). The acid process leads to denaturation and co-precipitation of heme proteins giving a less stable and darker product (Kristinsson *et al.* 2006).

Results on the influence of the pH shift process on functional properties of isolated fish protein are conflicting and depend both on species, condition of the raw material and processing conditions. The alkali process has been claimed to produce gels superior over both the acid-aided process and the surimi process. Sarcoplasmic proteins are removed from surimi because they are believed to interfere with gelation (Park *et al.* 1997). Other studies have on the other hand shown that the presence of sarcoplasmic protein had either no or a positive effect on gel strength (Kristinsson *et al.* 2006).

5.2 Fish protein hydrolysates

Fish protein hydrolysates (FPH) are produced industrially with the help of enzymes. They can be processed further by separating them into fractions using centrifugation, sieving, ultra- and nanofiltration and concentrating them by evaporation and drying. The products can be ingredients in fish and other food formulations, seafood flavours or ingredients for functional foods, nutraceuticals and cosmetics (Kristinsson 2006; Guerard 2006).

Mincing, homogenisation and mixing with enough water is the first step in the production of fish protein hydrolysates. The second step is the hydrolysis itself that includes setting the mixture to a right temperature, adjusting the pH and adding the enzyme. The choice of enzymes depends on the end product but also on cost. Commercial enzymes, usually a combination of endo- end exopeptidases are used in controlled hydrolysis (Guerard 2006) producing FPH with very different properties. Limited hydrolysis requires specific enzymes but a mixture of enzymes is used in more extensive hydrolysis. Extensive enzymatic hydrolysis of fish proteins generally produces protein hydrolysates with an amino acid composition similar to the fish protein (Kristinsson and Rasco 2002; Slizyte *et al.* 2005).

5.3 Fish sauce

Processing of fermented fish sauce is a traditional way of preserving fish in South-East Asia. Fish sauce is mainly produced from anchovies, mackerel and herring. Biochemically, fish sauce is salt-soluble protein in the form of amino acids and peptides. It is developed microbiologically with halophilic bacteria, which produce proteases that along with proteases from the fish muscle and viscera hydrolyse the fish and are principally responsible for flavour and aroma (Lopetcharat *et al.* 2001; Gildberg and Thongthai 2001; Fukami *et al.* 2004).

There is a great interest in producing low-salt fish sauce. The high-salt content inhibits the action of the enzymes in the fish which makes the processing time very long, and the great amount of salt in the product makes it difficult to market it as a health product. The processing time can be shortened by the addition of fish viscera or proteases (Kim *et al.* 1997; Morioka *et al.* 1999) or the reduction of salt concentration under 20% (Morioka *et al.* 1999; Gildberg and Thongthai 2001). It has also been documented that high pressure treatment under 60 MPa at 50 °C for 48 hours can be used to produce autolytate such as fish sauce without any addition of salt (Okasaki and other 2003). Fish sauce is mostly produced in South East Asia but successful production from Arctic species of pelagic fish like capelin has been reported (Gildberg 2001; Hjalmarsson and Kristbergsson 2007).

6. Isolation, purification and cleaning methods

6.1 Filtration

Peptides can be separated from hydrolysed and non-hydrolysed proteins with ultrafiltration membranes with molecular cut-off (MWCO) of 20 kDa or higher. The peptides in the permeate can then be fractionated with UF membranes of intermediate MWCO of approximately 4-8 kDa. Peptide solutions can also be concentrated with nanofiltration (NF) membranes of low MWCO, approximately 200-300 Da which will retain almost all the peptides except the smallest ones (Vandanjon *et al.* 2007). Ultrafiltration and nanofiltration can also be used in a diafiltration mode for desalination and partial deodorisation with nanofiltration membranes (Simon *et al.* 2002; Vandanjon *et al.* 2005).

6.2 Chromatography

There may be future opportunities in applying process chromatography to fish proteins and peptides. In a review article (Curling 2007), process chromatography is said not to have advanced as much as membrane separations but it is claimed to be on the threshold of industrialisation. It includes gel filtration to remove low molecular weight solutes, ion exchange chromatography to fractionate protein/peptide solutions, affinity chromatography to isolate and purify certain components. Ion exchange has been used in chromatographic processing of blood plasma fractionation.

7. Functional properties of marine proteins and peptides

Solubility, water-binding capacity, gelation and emulsifying properties are important functional properties of proteins used in food applications. They are influenced by protein source, production and environmental parameters. Production parameters are isolation, precipitation, drying or dehydration, concentration, modification (enzymatic, alkaline, acid hydrolysis, chemical) and environmental parameters include temperature, pH and ionic strength (Kinsella 1976).

Enzymatic hydrolysis changes the functional properties of fish proteins. Solubility is drastically improved, water-holding capacity remains fairly good and the emulsifying properties are reduced (Kristinsson and Rasco 2002; Slizyte *et al.* 2005). Heating is used to stop enzyme reaction and in the drying of FPH. It may have a deleterious effect on the functional properties and on nutritional and quality factors because of denaturation of proteins. Spray and freeze drying are considered a gentle form of drying and give a product where some retention of functional properties is observed. Drum drying can on the other hand cause losses in solubility.

Very little has been documented on the use of dried hydrolysed fish protein in injected or tumbled products. It has been reported that dried cod protein hydrolysates had better water-holding capacity than soy proteins but that the soy protein resulted in lower drip and higher cooking yield (Thorarinsdottir *et al.* 2004).

Intact fish myofibrillar proteins are quite insoluble in water over a wide pH range. Smaller peptides produced by hydrolysis have more of the hydrophilic polar amino acid side groups exposed and can bind more readily to water than the intact protein can (Kristinsson and Rasco 2000). High solubility over a wide range of pH is important for many food applications as it influences other functional properties, such as emulsifying and foaming properties. Good solubility of fish protein hydrolysates over a wide range of pH that increased with degree of hydrolysis has been reported several times (Sathivel *et al.* 2003; Gbobgouri *et al.* 2004; Sathivel *et al.* 2005; Sathivel and Bechtel 2006; Geirsdottir *et al.* 2007; Klompong *et al.* 2007; Thiansilakul *et al.* 2007).

Proteins are used as emulsifiers in foods because of their ability to produce desirable physicochemical properties in oil-in-water emulsions. Emulsifying properties of many fish proteins and fish protein hydrolysates have been reported. Interfacial activities (emulsion activity index, emulsion stability index, foaming capacity, foam stability) increase with limited hydrolysis but increased hydrolysis decreases them (Gbobgouri *et al.* 2004; Sathivel and Bechtel 2006; Sathivel *et al.* 2005; Klompong *et al.* 2007; Thiansilakul *et al.* 2007). FPH often show poorer emulsifying properties than dairy and soy proteins (Dickinson and Lopez 2001; Sathivel *et al.* 2003; Suhr *et al.* 2006). Good and similar properties have also been reported (Thiansilakul *et al.* 2007).Two protein fractions extracted from cod were able to form and stabilise oil-in-water emulsions (Petursson *et al.* 2004). Fraction between 10-30 kDa of cod frame protein hydrolysates showed excellent emulsion properties and whippability (Jeon *et al.* 1999).

8. Influence of process parameters on the bioactive, nutritional and sensory properties

8.1 Bioactive properties

Several *in vitro* and some *in vivo* trials that show bioactive properties of FPH have been documented and in some cases the active peptides have been isolated and identified.

Bioactive properties related to fish proteins are usually confined to short peptides released during digestion or hydrolysis. The bioactive properties of FPH are covered in another chapter in this book (Undeland *et al.* 2009). They have also recently been reviewed in connection with the Integrated Project SEAFOODplus (Guerard 2006; Thorkelsson *et al.* 2008). The release and amount of the peptides is influenced by the condition of the raw material, the hydrolysis process and concentration, fractionation and isolation processes. Fresh and clean raw material is the first criterion for production FPH with bioactive properties. Oxidation, changes in pH, freezing and freezer storage can change amino acids and reduce both solubility and digestibility. Bioactive peptides are formed during extensive hydrolysis using commercial mixtures of enzymes. The types of enzymes used and degree of hydrolysis depend on the intended bioactive property. Fermentation an also be used to produce both ACE-inhibiting and immunostimulating peptides. Reacting FPH with reducing sugars can increase antioxidant activity. Very short chain peptides with 2-8 amino acid residues isolated from FPH have shown the highest *in vitro* and *in vivo* ACE inhibiting activity. Longer peptides with 5-14 amino acid residues have shown *in vitro* antioxidative activities. Fractionation using ultrafiltration and nanofiltration are therefore used to concentrate the active peptides after hydrolysis.

8.2 Nutritional properties

Although extensive hydrolysis provides a pool of amino acids with a composition similar to the raw material, the nutritive quality is different. Certain qualities may be improved, whereas others may be reduced. Generally a mixture of free amino acids and small peptides are an excellent nitrogen source for microorganisms and infant individuals with a poorly developed digestive system. However, such mixtures do not provide optimal protein utilisation in adult individuals with a fully developed digestive system. This was revealed by experiments with young rats fed diets supplemented with minced fish and fish protein hydrolysates with different degree of hydrolysis (Espe *et al.* 1992). The young animals performed best on the moderately digested fish protein, but both protein efficiency ratio (PER) and weight gain were poorest with the extensively hydrolysed fish protein. The explanation for this was supposed to be that the absorption of amino acids is faster when a hydrolysed protein is ingested and that the influx of amino acids exceeds the capacity of the liver to synthesise proteins. In such situations, excess amino acids may be degraded or transformed into glycogen and lipid.

8.3 Sensory properties

The taste and odour of a protein hydrolysate depend on the raw material, the kind of protease applied and the hydrolytic conditions. Whereas high quality fish sauce provides a delicious flavour (Thongthai and Gildberg 2005), a major problem with most fish protein hydrolysates is the bitter taste (Kristinsson and Rasco 2002). The bitterness is normally caused by a number of medium size peptides with hydrophobic amino acid residues. Principally this problem may be solved either by performing mild hydrolysis to reduce production of medium size peptides or by running extensive hydrolysis to digest the troublesome peptides to free amino acids. Although mild hydrolysis may improve both flavour and nutritional properties, it will normally reduce the yield significantly. The latter is a big problem if maximal utilisation of the raw material is a major option. Extensive digestion is probably more convenient although it may reduce nutritional quality. Producers of commercial enzymes claim that the problem of bitterness may be solved by applying certain enzyme products with specific 'non-bitter' properties. However, this does not always hold true (Gildberg *et al.* 2002).

It is necessary to stabilise the residual lipid associated with the proteins by suitable processing techniques if fish proteins are to be used as food ingredients. Oxidation must be prevented during all processing steps. Off-flavour caused by protein and lipid oxidation can be a problem in dried fish proteins and fish protein hydrolysates. Mild processing conditions during hydrolysis and drying and lean raw material have been suggested to reduce the problem (Kristinsson and Rasco 2002). Oxidation is influenced by storage time and temperature. Other means include washing the lipids and heme proteins from the raw material, adding antioxidant before hydrolysis and selecting enzymes that do not operate at low pH and high temperatures (Kristinsson 2006). Advanced lipid oxidation in fresh spray-dried enzyme-hydrolysed saithe powder has been reported (Bragadottir *et al.* 2007). The drying of pH-shift processed protein could even exacerbate the problem (Geirsdottir 2006).

8.4 Non-proper compounds

Most commercial enzymes are of microbial origin, and some of them are quite thermostable. This underlines the importance of proper thermo inactivation to secure that no residual activity remains in the final product. Normally enzymatic protein hydrolysis is run under weak acid, neutral or weak alkaline conditions. If hydrolysis is performed under strong alkaline conditions the nutritional quality may be reduced due to racemisation of amino acids (De Boni and Scriba 2007), and even worse, formation of lysinoalanine which may have carcinogenic properties (Gilani *et al.* 2005). Strong acid conditions during hydrolysis may promote degradation of the essential amino acid tryptophan.

9. Fish protein and peptide products on the market

The market for marine protein and peptide products is not large compared with dairy and soy proteins and peptides. The traditional market for the products is in seafood flavours. The annual production of fish collagen/gelatine is about 3-6 thousand tonnes and accounts for only 1-2% of the global collagen/gelatine production. It is sold both for its physical and bioactive properties.

The market for functional foods and food supplements is growing very fast. There are many products with proteins and peptides mainly soy and dairy based. The number of collagen-based products is growing very fast. The number of collagen health foods and supplement products in Japan is remarkable (Functional Foods 2006). In Europe and the United States they are also sold as food supplements and to the cosmetic industry. Fish protein and peptide products with approved health claims do not exist in Europe and North America. There they are sold as food supplements.

In Table 3 there are examples of commercial products containing fish protein hydrolysates. Two products have been approved by Japanese authorities. One is the Katsuobushi oligopeptide made by hydrolysis of dried bonito. It is marketed as PEPIDE ACE 3000 in Japan (Functional Foods 2006). It is also marketed in the United States as Vasotensin® and PeptACE™ and Levenorm™ in Canada. The other is the sardine peptide SP100N a hydrolysed extract from sardine muscle and is among other products sold as a drink, LAPIS SUPPORT in Japan. Secure® a white fish protein hydrolysate concentrate has been on the market since 1994 in the United States. It is claimed to support the cells in your gastrointestinal tract and regulate bowel functions. Nutripeptin® is a peptide powder from codfish from Norway for reducing blood sugar.

Table 3. *Examples of commercially available functional foods or food ingredients carrying bioactive peptides.*

Product name	Claims	References
Hydrolysed dried bonito bowels		
Peptide ACE 3000	Lowers blood pressure	www.nippon-sapuri.com/english/
Vasotensin®	Lowers blood pressure	www.metagenics.com
PeptACE™	Lowers blood pressure	http://us.naturalfactors.com/
Levenorm™	Lowers blood pressure	www.onc.ca/
Peptides from sardines		
Lapis Support	Lowers blood pressure	http://www.tokiwayakuhin.jp/
Collagen peptides		
Bifidus & Collagen	Beautifies the skin	http://www.kagome.co.jp/
Hydrolysed whitefish		
Seacure	Improves gastrointestinal health	http://www.propernutrition.com/
Protizen	Relaxing	http://www.copalis.fr/
AntiSress 24	Relaxing	www.fortepharma.com/fr/index.html
Fortidium	Against oxidative stress	www.biothalassol.com/
Nutripeptin	Lowers glycaemic index	nutrimarine.com

10. Conclusions

The success of marketing of proteins in seafood or fish protein and peptide ingredients as functional foods or health food supplements depends, besides supplying substantial scientific evidence, on the taste and odour of the products. Good sensory quality of the products depends on simple and easily understandable criteria like fresh raw material, good handling and short processing time as well as on gaining more knowledge through research on the complicated interactions between process and environmental parameters in order to control oxidation and the formation of a bitter taste.

The pH-shift methods are being tested and are in use for increasing protein yield in traditional processing of lean white fish both in the United States and Europe. Applying protein isolates in seafood products can result in great additional economic, nutritional and environmental benefits. Much more research and development is needed into applying the isolates as commercial food ingredients. An even greater nutritional and economic advantage could be gained if pH-shift methods could be used to produce high quality isolates from fatty pelagic fish where there are storage and stability problems that need to be solved.

Fish protein hydrolysates will continue to be produced and sold as flavours and for infant feeds. The market based on special bioactive properties of fish protein hydrolysates is small and in the emerging phase of development. It has been forecast to grow annually about 8-12% until the year 2012 (Affertsholt 2007).

There are also opportunities in adapting traditional food processes like fermentation to increase the bioactive properties of fish protein hydrolysates and to employ them in products that consumers already know. Low-salt fish sauce and fish flavours with tailor-made bioactive properties are also likely products in the future.

Scientific documentation and official acceptance of health claims is needed if new functional products with fish proteins and peptides are to be marketed. Products based on soft generic nutritional claims are already on the market and it should be easy to develop more of them as much of the scientific documentation already exists. But it will be difficult and expensive to develop products claiming to reduce the risk of a disease. Here the scientific evidence is based on several *in vitro* studies and few *in vivo* studies. Collaboration is needed. Technological platforms for research and development and marketing to build up competence in marine ingredients are being created and built up in many places all over the world. Food ingredient companies that are very active in research and development in omega-3 fatty acids will also be involved in marine proteins and peptides in collaboration with both universities and private research organisations. But it will take years even decades for the seafood ingredients industry to be at the same level as the dairy industry is today in utilizing proteins and peptides from what was once called waste or by-products.

References

Affertsholt T. 2007. Trender i internasjonale markeder for ingredienser, fiskemel og fiskeolje. Biomarint industriseminar, Fiskeriforskning, Bergen 4-5.des. http://www.itek.norut.no/nofima/nyheter/seminar_og_konferanser/biomarint_industriseminar_2007

Anonymous. 2002. Regulation (EC) No 178/2002 laying down the general principles and requirements of food law, establishing the European Food Safety Authority and laying down procedures in matters of food safety.

Anonymous. 2004. Commission recommendation 2004/705/EC on the monitoring background levels of dioxins and dioxin like PCBs in foodstuffs.

Anonymous. 2005. Regulation (EC) No 2073/2005 on microbiological criteria for foodstuffs.

Aubourg SP, Medina I. 1999. Influence of storage time and temperature on lipid deterioration during cod (*Gadus morhua*) and haddock (*Melanogrammus aeglefinus*) frozen storage. Journal of the Science of Food and Agriculture 79(13):1943-1948.

Aubourg SP, Sotelo CG, Perez-Martin R. 1998. Assessment of quality changes in frozen sardine (*Sardina pilchardus*) by fluorescence detection. Journal of the American Oil Chemists Society 75(5):575-580.

Aubourg SP. 2001. Review: Loss of quality during the manufacture of canned fish products. Food Science and Technology International 7(3):199-215.

Baron CP, Andersen HJ. 2002. Myoglobin-induced lipid oxidation. A review. J Agric Food Chem 50(14):3887-97.

Batista I.1999. Recovery of proteins from fish waste products by alkaline extraction. European Food Research and Technology 210(2):84-89.

Bragadottir M, Reynisson E, Thorarinsdottir KA, Arason S. 2007. Stability of fish powder made from saithe (*Pollachius virens*) as measured by lipid oxidation and functional properties. Journal of Aquatic Food Product Technology 16(1):115-136.

Choe E, Min DB. 2006. Chemistry and reactions of reactive oxygen species in foods. Crit Rev Food Sci Nutr 46(1):1-22.

Curling J. 2007. History of chromatography: Process Chromatography: Five Decades of Innovation. BioPharm International Feb. Supplement.

Davies KJ. 1987. Protein damage and degradation by oxygen radicals. I. general aspects. J Biol Chem 262(20):9895-9901.

Davies MJ. 2005. The oxidative environment and protein damage. Biochim Biophys Acta 1703(2): 93-109.

De Boni S, Scriba GKE. 2007. Capillary electrophoresis analysis of hydrolysis, isomerization and enantiomerization of aspartyl model tripeptides in acidic and alkaline solution. J. Pharmaceut. Biomed. Analysis 43:49-56.

Dickinson E, Lopez G. 2001. Comparison of the emulsifying properties of fish gelatine and commercial milk proteins. J Food Sci 66(1):118-123.

Dissaraphong S, Benjakul S, Viessanguan W, Kishimura H. 2006. The influence of changes in fish sauce during fermentation. Bioresource Technology 97(16):2032-2040.

Emborg J, Dalgaard P, Ahrens, P. 2006. *Morganella psychrotolerans* sp. nov. a histamine-producing bacterium isolated from various seafoods. International Journal of Systematic and Evolutionary Microbiology 56:2473-2479.

Espe M, Haaland H, Njaa LR, Raa J. 1992. Growth of young rats on diets based on fish silage with different degrees of hydrolysis. Food Chem 44:195-200.

FAO. 2004. The State of World Fisheries and Aquaculture. Rome FAO.

FDA. 2004. Agency Response Letter 'GRAS Notice No.GRN 000147. www.cfsan.fda.gov/~rdb/opa-g147.html.

FIN. Fishmeal Information Network. 2007. 'Fishmeal and Fish oil facts and figures', http://www.gafta.com/fin/pdfs/FACTSANDFIGURES.pdf

Fukami K, Funatsu Y, Kawasaki K, Watabe S. 2004. Improvement of fish sauce odor by treatment with bacteria isolated from the fish sauce mush (Moromi) made from frigate mackerel. J Food Sci 69(1):45-49.

Functional Foods. Japan 2006. Product Report' New York, Paul Yamaguchi and Associates Inc, www.functionalfoodjapan.com

Gbobgouri GA, Linder M, Fanni J, Parmentier M. 2004. Influence of hydrolysis degree on the functional properties of salmon by-products hydrolysates. J Food Sci 69(8):615-622.

Geirsdottir M, Hlynsdottir H, Thorkelsson G, Sigurgisladottir S. 2007. Solubility and viscosity of herring (*Clupea harengus*) proteins as affected by freezing and frozen storage. J Food Sci 72(7):376-380.

Geirsdottir M. 2006. Protein isolation from herring. IFL Report 39-06.

Gilani GS, Cockell KA, Sepehr E. 2005. Effects of antinutritional factors on protein digestibility and amino acid availability in foods. J AOAC Internat 88:967-987.

Gildberg A, Arnesen JA, Carlehög M. 2002. Utilisation of cod backbone by biochemical fractionation. Process Biochem 38:475-480.

Gildberg A, Thongthai C. 2001. The effect of reduced salt content and addition of halophilic lactic bacteria on quality and composition of fish sauce made from sprat. Journal of Aquatic Food Product Technology 10:77-88.

Gildberg A. 2001. Utilisation of male Arctic capelin and Atlantic cod intestines for fish sauce production-evaluation of fermentation conditions. Bioresource Technology 76(2):119-123.

Guerard F 2006. Enzymatic methods for marine by-product recovery. In: Shahidi F (ed.) Maximising the value of marine by-products, Cambridge: Woodhead Publishing Limited. p. 107-143.

Guerard F, Sumaya-Martinez MT, Laroque D, Chabeaud A, Dufossé L. 2007.Optimization of free radical scavenging activity by response surface methodology in the hydrolysis of shrimp processing discards. Process Biochemistry 42(11):1486-1491.

Hidalgo FJ, Zamora R. 2002. Methyl linoleate oxidation in the presence of bovine serum albumin. J Agric Food Chem 50(19):5463-5467.

Hjalmarsson G H, Park JW, Kristbergsson K. 2007. Seasonal effects on the physicochemical characteristics of fish sauce made from capelin (*Mallotus villosus*). Food Chemistry 103(2):495-504.

Hultin HO, Kelleher SD, Feng Y, Kristinsson HG, Undeland I, Ke S. 2004. High Efficiency Alkaline Protein Extraction. US Patent Application No.10/363.612.

Hultin HO, Kelleher SD. 1999. Process for isolating a protein composition from a muscle source and protein composition. U.S Patent No. 6,005,073.

Hultin HO, Kristinsson HG, Lanier TTC, Park JW. 2005. Process for recovery of functional proteins by pH shifts. In: Park JW (ed.) Surimi and Surimi Seafood. 2nd Eds, New York: Marcel Dekker. p. 107-139.

Huss HH. 1995. Quality and changes in fresh fish.FAO Fisheries Technical Paper -348. Food and Agriculture Organization of the United Nations. M-47 ISBN 92-5-103507-5.

Jeon YJ, Byun HG, Kim SK. 1999. Improvement of functional properties of cod frame protein hydrolysates using ultrafiltration membranes. Process Biochemistry 35(5): 471-478.

Kelleher SD, Williamson PG. 2007. Food product and process for reducing oil and fat content in cooked food. US patent 7,163,707.

Kim SK, Jeon YJ, Byun HG, Kim YT, Lee CK. 1997. Enzymatic recovery of cod frame proteins with crude proteinase from tuna pyloric caeca. Fisheries Sci 63(3):421-427.

Kinsella JE. 1976. Functional properties of proteins in foods: A survey. Critical Review in Food Science and Nutrition 7(3):219-280.

Klompong V, Benjakul S, Kantachote D, Shahidi F. 2007. Antioxidative activity and functional properties of protein hydrolysate of yellow stripe trevally (*Selaroides leptolepis*) as influenced by the degree of hydrolysis and enzyme type. Food Chemistry 102(4):1317-1327.

Kristinsson HG, Hultin HO. 2004. Changes in trout hemoglobin conformations and solubility after exposure to acid and alkali pH. J Agric Food Chem 52(11):3633-3643.

Kristinsson HG. and Rasco BA. 2002. Fish protein hydrolysates and their potential use in the food industry. In: Fingerman M and Nagabhushanan R (eds.) Recent Adv. In Marine Biotechnol., Vol. 7, Sci. Publ. Inc., Enfield, NH. p. 157-181.

Kristinsson HG, Rasco BA. 2000. Fish Protein Hydrolysates: Production, Biochemical, and Functional Properties. Critical Reviews in Food Science and Nutrition 40(1):43-81.

Kristinsson HG, Theodore AE, Ingadottir B. 2006. Chemical processing methods for protein recovery from marine by-products and underutilized species. In: Shahidi F (ed.) Maximising the value of marine by-products. Cambridge:Woodhead Publishing Limited. p. 144-168.

Kristinsson HG. 2002. Acid-induced unfolding of flounder haemoglobin: Evidence for a molten globular state with enhanced pro-oxidative activity. J Agric Food Chem 50(26):7669-7676.

Kristinsson HG. 2006. Aquatic food protein hydrolysates. In: Shahidi F (ed.) Maximising the value of marine by-products. Cambridge: Woodhead Publishing Limited. p. 229-247.

Li SJ, King AJ. 1999. Structural changes of rabbit myosin subfragment 1 altered by malonaldehyde, a byproduct of lipid oxidation. J Agric Food Chem 47(8):3124-3129.

Lopetcharat K, Chai YJ, Park JW, Daeschel M. 2001. Fish sauce products and manufacturing: a review. Food Rev. International 17(1):65-88.

Morioka K, Fujii S, Iton Y, Chengchu L, Obatake A. 1999. Recovery of amino acid from protein in the head and viscera of frigate mackerel by autolysis. Fisheries Science 65(4):588-591.

Okasaki T, Shigeta Y, Aoyama Y, Mamba K. 2003. Autolysis of unsalted fish protein under pressurization', Fisheries Science 69(6):1257-1263.

Park JW, Lin TM, Yongsawatatdigul J. 1997. New developments in manufacturing of surimi and surimi seafood. Food Rev International 13(4):577-610.

Petursson S, Decker EA, McClements DJ, 2004. Stabilization of oil in water emulsions by cod protein extracts. J Agric Food Chem 52(12):3996-4001.

Rampon V, Lethuaut L, Mouhous-Riou N, Genot C. 2001. Interface characterization and aging of bovine serum albumin stabilized oil-in-water emulsions as revealed by front-surface fluorescence. Journal of Agricultural and Food Chemistry 49(8):4046-4051.

Refsgaard HH, Tsai L, Stadtman, ER. 2000. Modifications of proteins by polyunsaturated fatty acid peroxidation products. Proc Natl Acad Sci USA 97(2):611-616.

Rustad T (2003). Utilisation of marine by-products. Electronic Journal of Environmental Agriculture and Food Chemistry 2:458-463.

Saeed S, Howell NK. 2002. Effect of lipid oxidation and frozen storage on muscle proteins of Atlantic mackerel (*Scomber scombrus*). Journal of the Science of Food and Agriculture 82(5):579-586.

Sathivel S, Bechtel PJ, Babbit B, Smiley S, Crapo C, Reppond KD, Prinyawiwatkul W. 2003. Biochemical and functional properties of herring (*Clupea harengus*), by-product hydrolysates. J Food Sci 68(7):2196-2200.

Sathivel S, Bechtel PJ. 2006. Properties of soluble protein powders from Alaska pollock (*Theragra chalcogramma*). Int J Food Sci Technol 41(5):520-529.

Sathivel S, Smiley S, Prinyawiwatkul W, Bechtel PJ. 2005. Functional and nutritional properties of red salmon (*Oncorhynchus nerka*) enzymatic hydrolysates. J Food Sci 70(6):C401-C408.

Shahidi F. 2006. Maximising the value of marine by-products, Cambridge: Woodhead Publishing Limited, 560 pp.

Simon A, Vandanjon L, Levesque G, Bourseau P. 2002. Concentration and desalination of fish gelatine by ultrafiltration and continuous diafiltration processes. Desalination 144(1):313-318.

Slizyte R, Dauskas E, Falch E, Storrø I, Rustad T. 2005. Characteristics of protein fractions generated from hydrolysed cod *(Gadus morhua)* by-products. Process Biochem 40:2021-2033.

Soyer A, Hultin HO. 2000. Kinetics of oxidation of the lipids and proteins of cod sarcoplasmic reticulum. Journal of Agricultural and Food Chemistry 48(6):2127-2134.

Suhr J, Decker EA, McClements DJ. 2006. Properties and stability of oil-in-water emulsions stabilized by fish gelatine. Food Hydrocolloids 20(5):596-606.

Thiansilakul Y, Benjakul S, Shahidi F. 2007. Compositions, functional properties and antioxidative activity of protein hydrolysates prepared from round scad (*Decapterus maruadsi*). Food Chemistry 103(4):1385-1394.

Thongthai C, Gildberg A. 2005. Asian fish sauce as a source of nutrition. In: Shi J, Ho CT, Shahidi F (eds.) Asian Functional Foods. Boca Raton: CRC Press. p. 215-265.

Thorarinsdottir KA, Gudmundsdottir G, Arason S, Thorkelsson G, Kristbergsson K. 2004. Effects of added salt, phosphates, and proteins on the chemical and physicochemical characteristics of frozen cod (*Gadus morhua*) fillets. J Food Sci 69(4):E144-E154.

Thorkelsson G, Sigurgisladottir S, Geirsdottir M, Jóhannsson R, Guérard F, Chabeaud A, Bourseau P, Vandanjon L, Jaouen P, Chaplain-Derouiniot M, Fouchereau-Peron M, Martinez-Alvarez O, Le Gal Y, Ravallec-Ple, Picot L, Berge JP, Delannoy IC, Jakobsen G, Johansson I, Batista I, Pires C. 2008. Mild processing techniques and development of functional marine protein and peptide ingredients. In: Børresen T (ed.) Improving seafood products for the consumer. Cambridge: Woodhead Publishing Limited. p. 363-398.

Thorsdottir I, Ramel A. 2008. Fish consumption and the health of children and young adults. In: Børresen T (ed.) Improving seafood products for the consumer. Cambridge: Woodhead Publishing Limited. 136-164

Tironi VA, Tomas MC, Anon MC. 2002. Structural and functional changes in myofibrillar proteins of sea salmon (*Pseudopercis semifasciata*) by interaction with malonaldehyde (RI). Journal of Food Science 67(3):930-935.

Tunhun D, Itoh Y, Morioka K, Kubota S, Obatake A. 2002. Gel forming ability of fish meat oxidized during washing. Fisheries Science 68(3):662-671.

Undeland I, Lindqvist H, Yun CY, Falch E, Ramel A, Cooper M, Gildberg A, Luten J, Stenberg E, Nielsen HH, Elvevoll E. 2009. Seafood and health: what is the full story? In: Luten JB (ed.) Marine functional foods. Wageningen, Wageningen Academic Publishers, pp. 17-87.

Vandanjon L, Bourseau P, Jaouen P. 2005. Concentration de solutions de peptides d'origine marine par ultrafiltration et nanofiltration. SFGP Congress. p. 20-22.

Vandanjon L, Johannsson R, Derouiniot M, Bourseau P, Jaouen P. 2007. Concentration and purification of blue whiting peptide hydrolysates by membrane processes. J Food Eng 83(4):581-589.

Vidal-Giraud B, Chateau D. 2007. World Surimi Market. Global Research Programme, Volume 89.

Xiong YLL. 2005. Role of myofibrillar proteins in water binding in brine-enhanced meats. Food Res Int 38(3):281-287.

Yong SH, Karel M. 1978. Reactions between peroxidizing lipids and histidyl residue analogues: enhancement of lipid oxidation and browning by 4-methylimidazole. Lipids 13(1):1-5.

Probiotics and seafood

Łukasz Grześkowiak, Jaakko Korpela and Seppo Salminen
University of Turku, Functional Food Forum, 20014 Turku, Finland

1. General health effects from probiotics

1.1 How to define probiotics?

According to the definition of the FAO/WHO working group, probiotics are 'live microorganisms which when administered in adequate amounts confer a health benefit on the host' (FAO/WHO 2002). The basic idea of orally administered probiotic is to affect the host's normal intestinal microbiota and change its composition from potentially harmful composition into a microbiota that would be beneficial for the host. Specific health effects are unique for each probiotic strain or product and they must be scientifically verified by clinical studies. Most common probiotic strains are selected from the genera *Lactobacillus* and *Bifidobacterium,* often referred to as lactic acid bacteria (Salminen *et al.* 1999; Kligler and Cohrssen 2008).

1.2 Documented and proposed health effects of probiotics

Multiple reports have described the beneficial effects of adding selected probiotic strains to food products. These health effects range from rather vague, e.g. increase in well-being, to more specific benefits, such as antagonism to certain pathogens (Balcázar *et al.* 2007). Some of these benefits are already well documented, but the majority of them are as yet only proposals based on limited scientific evidence.

The strongest evidence of health effects is related to the use of probiotics in prevention and treatment of acute rotavirus diarrhea and the ability of certain probiotic strains to improve lactose tolerance in lactose-deficient persons. Other clinically well-documented effects are the prevention of urinary tract infections and the management of antibiotic-associated diarrhea. Other diarrheal problems, such as *Clostridium difficile*-associated, traveller's and radiotherapy-associated diarrhea are promising, but less well established. Promising results have also been obtained in the blood pressure lowering effect of probiotics (Goossens *et al.* 2003; Gill and Guarner 2004; Sullivan and Nord 2005; Vanderhoof 2008).

A few other gastrointestinal disorders may be effectively treated or prevented with probiotics. Recent studies have documented effects in irritable bowel syndrome and different inflammatory bowel diseases (Fedorak 2008; Heilpern and Szilagyi 2008; Mengheri 2008) as well as in short bowel syndrome, constipation and necrotizing enterocolitis of preterm neonates. Many reports describe the use of probiotics in attempts to eradicate *Helicobacter pylori* inflammations, but probiotics seem only to improve the success rates with standard treatments in an adjunct manner (Kim *et al.* 2008; Scaccianoce *et al.* 2008; Sachdeva and Nagpal 2009).

In addition to the overall effects on well-being and health, probiotic microbes have many specific benefits via influencing human immune system. Probiotics can modulate the immune function and alter inflammatory processes which may have an impact on many inflammatory diseases outside the intestinal tract as well (Heilpern and Szilagyi 2008). There are some results which strongly indicate that early age administration of certain probiotics can reduce the prevalence of atopic diseases in infants. Some preliminary studies have already proposed probiotic treatments for some other allergies, e.g. food allergies, pollen allergies and rhinitis (Betsi 2008; Schabussova and Wiedermann 2008).

Despite the lack of clinical evidence for the anticancer action of probiotics, there are indirect proofs of probiotic microbes reducing the cancer risk. This may be a result of removing the dietary carcinogens/mutagens or decreasing the faecal enzymes. Another approach is their production of protective metabolites, such as antioxidants or short-chain fatty acids (De Vrese and Schrezenmeir 2008; Douglas and Sanders 2008; Kligler and Cohrssen 2008).

Less documented health effects of probiotics include a reduction in serum cholesterol levels and a reduction in the risk of diabetes (Ljungberg *et al.* 2006; De Vrese and Schrezenmeir 2008; Douglas and Sanders 2008). There is a recent clinical study indicating that deviated gastrointestinal microbiota may be an additional factor enhancing obesity development by increased the capacity to harvest energy from diet (Turnbaugh *et al.* 2006). It has also been suggested that abnormal microbiota may play a role in certain psychological conditions, such as clinical depression and onset of autism. In the latter case, probiotics are already being used as an adjuvant treatment (Schabussova and Wiedermann 2008).

Taken together, there are few scientifically well-validated effects on human health but also a number of challenging opportunities for new probiotics as well as new target populations with specific microbiota aberrancies.

2. Probiotic bacteria in seafoods

Probiotic microorganisms are increasingly used in food products due to their proven health benefits for human (Heller 2001). A common example of such functional food is fermented milk containing live probiotic strain(s). The use of lactic acid bacteria and their bacteriocins as biopreservatives also has potential for application in seafood products.

The object of this short contribution was to find (commercial) fish or marine foods, in which probiotic cultures are either added or used in manufacture. So far no such product has been found. Most related products are traditional, naturally preserved fish dishes from South-East Asia, Africa and also Scandinavia (Tanasupawat and Komagata 1995; Kobayashi 2000). Scientific reports of these seafoods concentrate mostly on consumer safety and nutritional aspects. Only a few papers also describe the biological preservation (biopreservation) of seafood with lactic acid bacteria (LAB).

In one study, lactic acid bacteria were isolated from fish intestine, smoked and marinated fish and they included strains of carnobacteria, lactococci and enterococci. Among them, two bacteriocin producing carnobacteria were further studied and they showed potential for use in fish preservation (Pilet 1995). Another report describes biopreservation of brined shrimp which is a highly perishable product that is normally preserved with added sorbic or benzoic acid. Three bacteriocins, Nisin Z, carnocin UI49 or bavaricin A were added to shrimp. Nisin Z was the most effective of these three bacteriocins, extending the storage life 21 days for the samples preserved with this bacteriocin (Einarsson and Lauzon 1995). Such bacteriocins can also be produced when specific lactic acid bacteria are added into the product.

Probiotics are already used as a feed additive in fish farming, but probiotic-containing fish products for human consumption do not seem to be available on the market.

In South East Asia, especially Thailand, fermenting is a traditional method for preserving different seafood such as fish sauce or fish paste. Fermentation occurs naturally, and starter cultures are rarely used. In low-salt (<8%) Thai fermented fish, souring and ripening is a result of growth and metabolism of a mixture of lactic acid bacteria. In contrast, in high-salt Thai fermented fish (>8%), only the salt itself is crucial for preventing spoilage (Tanasupawat and Komagata 1995). These foods probably have a healthy role as a local cuisine, but specific health effects cannot be established since the fermentation process is spontaneous and no probiotic starter strains are used. Traditional Swedish 'surströmming' is spontaneously fermented and canned herring filets, whose unique flavour is mostly due to halophilic bacteria (Kobayashi *et al.* 2000).

'Biopreservation' refers to the extended storage life and enhanced safety of foods using the natural or controlled microflora and/or their antibacterial products (Stiles 1996). Recent studies have demonstrated that inoculated LAB inhibited the growth of unwanted food spoilage microbes, especially *Listeria* spp. in cold smoked salmon (Leroi *et al.* 1996; Tome *et al.* 2006; Weiss and Hammes 2006).

Apart from seafood, probiotic cultures of LAB have been used in dry meat sausage fermentation. Clinical evidence of such meat products are minimal, but the combined health benefits and protective nature of such probiotic strains seems promising (Tyopponen 2003; Leroy *et al.* 2006). These products are commercially available in some European countries.

In conclusion, there is a potential for specific selected probiotic bacteria both in fish farming and preservation of fish products with possibilities of extending the probiotic properties to the end-user.

References

Balcázar JL, Vendrell D, De Blas I, Ruiz-Zarzuela I, Gironés O, Muzquiz JL. 2007. *In vitro* competitive adhesion and production of antagonistic compounds by lactic acid bacteria against fish pathogens. Veterinary Microbiology 122:373-380.

Betsi GI, Papadavid E, Falagas ME. 2008. Probiotics for the treatment or prevention of atopic dermatitis: a review of the evidence from randomized controlled trials, Am J Clin Dermatol. 9(2):93-103.

De Vrese M, Schrezenmeir J. 2008. Probiotics, prebiotics, and synbiotics, Adv Biochem Eng Biotechnol. 111:1-66.

Douglas LC, Sanders ME. 2008. Probiotics and prebiotics in dietetics practice, J Am Diet Assoc. 108(3):510-521.

Einarsson H, Lauzon HL. 1995. Biopreservation of brined shrimp (Pandalus borealis) by bacteriocins from lactic acid bacteria. Appl. Environ. Microbiol. 61:669-676.

FAO/WHO, 2002. Guidelines for the evaluation of probiotics in food. Food and Agricultural Organization of the United Nations and World Health Organization. Working Group Report.

Fedorak RN. 2008. Understanding why probiotic therapies can be effective in treating IBD, J Clin Gastroenterol. 42 Suppl 3 Pt 1:S111-S1115.

Gill HS, Guarner F. 2004. Probiotics and human health: a clinical perspective, Postgrad Med J. 80(947):516-526.

Goossens D, Jonkers D, Stobberingh E, Van den Bogaard A, Russel M, Stockbrügger R. 2003. Probiotics in gastroenterology: indications and future perspectives, Scand J Gastroenterol Suppl. 239:15-23.

Heilpern D, Szilagyi A. 2008. Manipulation of intestinal microbial flora for therapeutic benefit in inflammatory bowel diseases: review of clinical trials of probiotics, pre-biotics and synbiotics, Rev Recent Clin Trials. 3(3):167-184.

Heller KJ. 2001. Probiotic bacteria in fermented foods: product characteristics and starter organisms, Am J Clin Nutr. 73:374S-379S.

Kim MN, Kim N, Lee SH, Park YS, Hwang JH, Kim JW, Jeong SH, Lee DH, Kim JS, Jung HC, Song IS. 2008. The effects of probiotics on PPI-triple therapy for Helicobacter pylori eradication, Helicobacter. 13(4):261-268.

Kligler B, Cohrssen A. 2008. Probiotics, Am Fam Physician. 78:1073-1078.

Kobayashi T, Kimura B, Fujii T. 2000. Strictly anaerobic halophiles isolated from canned Swedish fermented herrings (Surstromming). Int J Food Microbiol. 54(1-2):81-89.

Leroi F, Arbey N, Joffraud J-J, Chevalier F. 1996. Effect of inoculation with lactic acid bacteria on extending the shelf-life of vacuum-packed cold smoked salmon. International Journal of Food Science and Technology. 31:497-504.

Leroy F, Verluyten J, De Vuyst L. 2006. Functional meat starter cultures for improved sausage fermentation. Int J Food Microbiol. 106(3):270-285.

Ljungberg M, Korpela R, Ilonen J, Ludvigsson J, Vaarala O. 2006. Probiotics for the prevention of beta cell autoimmunity in children at genetic risk of type 1 diabetes--the PRODIA study, Ann N Y Acad Sci. 1079:360-364.

Mengheri E. 2008. Health, probiotics, and inflammation, J Clin Gastroenterol. 42 Suppl 3 Pt 2:S177-S178.

Pilet ME, Dousset X, Barry R, Novel G, Desmazeaud M, Piard JC. 1995. Evidence for two bacteriocins produced by *Carnobacterium piscicola* and *Carnobacterium divergens* isolated from fish and active against *Listeria monocytogenes*. J. Food Prot. 58:256-262.

Sachdeva A, Nagpal J. 2009. Effect of fermented milk-based probiotic preparations on *Helicobacter pylori* eradication: a systematic review and meta-analysis of randomized-controlled trials, Eur J Gastroenterol Hepatol. 21(1):45-53.

Salminen S, Ouwehand A, Benno Y, Lee YK. 1999. Probiotics: how should they be defined? Trends in Food Science and Technology. 10:107-110.

Scaccianoce G, Zullo A, Hassan C, Gentili F, Cristofari F, Cardinale V, Gigliotti F, Piglionica D, Morini S. 2008. Triple therapies plus different probiotics for *Helicobacter pylori* eradication, Eur Rev Med Pharmacol Sci. 12(4):251-256.

Schabussova I, Wiedermann U. 2008. Lactic acid bacteria as novel adjuvant systems for prevention and treatment of atopic diseases, Curr Opin Allergy Clin Immunol. 8(6):557-564.

Stiles, ME. 1996. Biopreservation by lactic acid bacteria. Antonie Van Leeuwenhoek. 70(2-4):331-345.

Sullivan A, Nord CE. 2005. Probiotics and gastrointestinal diseases, J Intern Med. 257(1):78-92.

Tanasupawat S, Komagata K. 1995. Lactic acid bacteria in fermented foods in Thailand. World Journal of Microbiology & Biotechnology. 11:253-256.

Tome E, Teixeira P, Gibbs PA. 2006. Anti-listerial inhibitory lactic acid bacteria isolated from commercial cold smoked salmon. Food Microbiol. 23(4):399-405.

Turnbaugh PJ, Ley RE, Mahowald MA, Magrini V, Mardis ER, Gordon JI. 2006. An obesity-associated gut microbiome with increased capacity for energy harvest. Nature. 21:444.

Tyopponen S, Petaja E, Mattila-Sandholm T. 2003. Bioprotectives and probiotics for dry sausages. Int J Food Microbiol. 83(3):233-244.

Vanderhoof JA. 2008. Probiotics in allergy management, J Pediatr Gastroenterol Nutr. 47 Suppl 2:S38-S40.

Weiss A, Hammes WP. 2006. Lactic acid bacteria as protective cultures against *Listeria* spp. on cold-smoked salmon. Eur Food Res Technol. 222:343-346.

Consumer acceptance of (marine) functional food

Pirjo Honkanen
Nofima Market, P.O. Box 6122, 9291 Tromsø, Norway

1. Introduction

Scientific and technological developments in the food sector have led to a marked shift in the way consumers deal with food and health. There is a growing awareness that the dietary source and form of food may affect the overall health of the consumer. The role of food as an agent for improving health has initiated the development of new classes of food - functional foods. Nutrients and other bioactive substances isolated from fish as well as fish itself can be used as ingredients for functional foods. There is no officially approved definition of functional foods (Lähteenmäki *et al.* 2007). We choose to follow the definition of Lähteenmäki *et al.* (2007), that functional foods are foods that are marketed with a health claim, that is, they promise some health-related benefit for the consumers. This excludes foods with modified nutrient content (e.g. low in fat, low in sodium, etc.) and dietary supplements (omega-3 capsules, etc.), which are more like medicine than food.

Motives for choosing functional foods are most likely staying healthy, maintaining general well-being, or/and preventing diseases. Europeans consider health to be one of the most important factors influencing their food choice, but a majority also think that they already have a healthy diet (Lappalainen *et al.* 1998). This may be a barrier to buying healthy and functional foods. Another factor which may limit the success of functional foods is that diet changes and nutrient intake have mainly long-term effects, which cannot be observed immediately. Few people seem to be concerned with the long-term effects of food on their health (Frewer *et al.* 2003). This may be a result of optimistic bias or deterministic beliefs in general. Optimistic bias implies that people think that serious health problems will not affect them to the same degree as it does others. The deterministic world view could cause people to disregard nutritional advice and healthy foods because they believe that it is not possible to influence one's health anyway, either by thinking that other factors are more important or just believing that one's health status is determined already. General awareness of the link between health and food does not seem to be sufficient to make people choose functional food (Urala 2005). There also seems to be a need to feel personal relevance (Van Kleef *et al.* 2005): for example, a sick family member (Verbeke 2005).

Sensory appeal, health, price and convenience are the most important motives for food choice in general (Steptoe *et al.* 1995). The motives for choosing functional food seem to be quite similar: healthiness, taste, pleasantness, convenience, price, familiarity and confidence (Bhaskaran and Hardley 2002; Urala and Lähteenmäki 2003). These reasons are product dependent, that is, the reasons can vary in importance from product to product.

We know very little about consumer acceptance of *marine* functional food. The clinical/medical studies about the health effects of seafood are documented in other sections in this book. There are many general studies about consumption of and attitudes towards seafood, which have found that fish is considered healthy food, but other foods are often preferred to fish. Such studies do not, however, take a closer look at consumer perceptions of the relationship between health and seafood. Due to the fact that literature on consumer acceptance of marine functional food is almost non-existent, we have chosen to review consumer acceptance of functional foods in general. We also suggest the need for further studies on marine functional foods, highlighting some potential differences in consumer acceptance compared to other functional foods.

There are many factors that can influence the consumer acceptance of functional food. We have chosen to concentrate on factors related to the consumer, the health claims and the product. Attitudes towards functional foods are described in section 1, consumers' knowledge and awareness of functional foods are examined in section 2. Section 3 describes the type of health claims and consumer trust in health claims, and in section 4 type of carrier products are reviewed (including perceived healthiness and naturalness and hedonic aspects of the product). In section 5, demographic and cross-cultural differences in the acceptance of functional foods are presented. Finally, in section 6, we draw some conclusions, dealing especially with issues related to marine functional food.

2. Attitudes

Attitudes towards food and food consumption are important predictors of food intake in general. This is assumed to hold for functional foods as well, and many articles have been published about consumers' beliefs and attitudes towards functional foods. The attitudes towards functional foods have been found to be positive in Finland, while the consumers in Denmark seem to be more sceptical (Poulsen 1999; Urala *et al.* 2003). In a study in Finland, Urala and Lähteenmäki (2007) found four dimensions that reflect consumers' attitudes towards and willingness to use functional foods: reward from using functional foods, necessity for functional foods, confidence in functional foods and safety of functional foods. The *reward* dimension is concerned with consumers' perception that health and general well-being can be promoted by consuming functional foods. The perceived reward was found to be the strongest predictor of acceptance. *Necessity* of functional foods is about how necessary it is to consume functional food in order to stay healthy.

Confidence describes the attitudes towards confidence in the health-promoting status of functional foods and the research related to them. The *safety* dimension describes the attitudes towards risks and doubts concerning functional foods. This dimension reflects the consumers' uncertainties about how much of a certain functional food they can eat without getting an overdose, for example. Niva (2007) also found that people are sceptical as to whether functional food is safe to eat: some think that it is a sham while others think it may actually be unhealthy. Urala (2005) found that the weight of the four attitude dimensions differed between different types of functional foods. All in all, it seems that the subject of functional food is still quite new to many people and the attitudes are not very stable, depending on familiarity with the product.

Risk perception can be important in the acceptance of functional foods. Just relating a benefit to novel foods may not be sufficient to gain consumer acceptance: understanding consumers' risk perceptions associated with processing technologies and their health status can be important as well (Frewer *et al.* 2003). Perceived benefits have to offset the negative associations with new processing methods.

There is some evidence that people may be sceptical about functional food as a therapeutic or dietary supplement (Bhaskaran and Hardley 2002). Several issues seem to emerge: how much and how frequently should the food be consumed? What is the correct dose? These concerns seem to reflect the difficulty in distinguishing between food and medicine.

2.1 Attitudes towards marine functional foods

In one of the few studies involving marine ingredients, Bech-Larsen and Grunert (2003) included an omega-3 enriched product in their con-joint study about perceived healthiness of functional foods in 2003. Their results indicated that some products might be more appropriate as carrier products compared to others: for example, juice and omega-3 was not a good combination.

Verbeke *et al.* (2005) studied consumer perceptions of fish that might be considered as functional food itself. It contains a lot of important nutrients (e.g. omega-3, vitamin D, iodine), which have a beneficial influence on health. The study showed that consumers believe that fish is healthy and nutritious. They also believe that fish reduces the risk of coronary heart disease, but the awareness of the specific nutrients is rather low. Consumers generally have quite positive attitudes towards fish and fish consumption, and the attitudes are often more positive the older the respondents are (Honkanen *et al.* 2005; Olsen 2003; Olsen *et al.* 2007).

3. Knowledge/awareness

Knowledge is found to be a major motivational factor for purchasing functional foods (Frewer *et al.* 2003), while lack of knowledge is a major reason not to consume functional foods (IFIC 1999). Both subjective and objective knowledge matter in addition to awareness of the relationship between the functional substance and health benefits.

Lack of basic nutrition knowledge has been found to limit consumers' ability to evaluate health claims (Williams 2006). Verbeke (2005) found in a study from 2001 in Belgium, that subjective knowledge was an important factor for functional food acceptance at the category level, but this effect was negative (that is, the more perceived knowledge the consumer has, the lower the intention to consume functional food), contradicting earlier findings. The results also showed that the negative effect was less important among the older respondents. The negative finding may, however, reflect the low level of factual knowledge among the consumers. Another study showed that the consumers' awareness and knowledge about the nutrients in fish is quite low (Verbeke *et al.* 2005). Most people know that fish contains a lot of omega-3 fatty acids, but they do not necessarily relate it to a positive effect on human health. Several other studies indicate that consumers in many countries have little knowledge about functional foods or how specific enrichments are related to health (Poulsen 1999).

Even though knowledge is important in understanding health claims and for the acceptance of functional foods, it may not be the key element. Most Americans, for example are aware of the relationship between omega-3 fatty acids and the reduced risk of heart disease (IFIC 2005), but they still haven't increased their seafood consumption. This seems to be the case also in Europe. The knowledge about the health effects of fish is high but there is no evidence of a substantial increase in consumption. This underlines the importance of studies looking at several factors working together in the acceptance of functional foods.

It seems that the findings on the impact of knowledge on the acceptance of functional food are somewhat conflicting, so further research is needed. This is especially important because knowledge level is likely to have an influence on what type of health claims should be used. People are not necessarily able to process messages with specific health/enrichment information.

4. Health claims: types and consumer trust

The type of health claim seems to have an influence on consumer acceptance of functional food. The use of health claims is, however, regulated by the authorities in many countries. In the US, nutritional claims can be used without approval (Van Trijp and Van der Lans 2007). However, claims that suggest curing, treating or prevention of diseases are not allowed in Europe (Urala and other 2003). In Europe, the regulation has until recently varied from country to country, but in 2006, a new EU regulation was passed. The EU regulation (No 1924/2006) lays down provisions for the use of health claims in labelling and advertising of food products. These provisions are highly important for functional foods, as they regulate what is or is not allowed in profiling such products. One important aspect is that any health or nutritional claim has to be backed up by scientific evidence, and the product should have a beneficial nutritional or physiological effect. Basically, unhealthy foods cannot be accompanied by health claims according to the new regulation. The directive differentiates between nutrition claims and health claims. Nutrition claims are only allowed if they are listed in the Annex attached to the directive. These are claims that state a beneficial nutritional property, for example, low in salt, low in fat, high fibre, etc. The literature about nutrition claims will not be reported here, because these claims do not state a health effect of the nutrient, and thus are not included in our definition of functional food.

Health claims can only be authorised for use after a scientific assessment of the claims by the EFSA. Health claims can mean any health claims that suggest a relationship between a food and health. The directive divides health claims into two categories:

Reduction of disease risk claim is a claim suggesting that the consumption of a food significantly reduces the risk factor in the development of a human disease. These claims have to be based on general scientific data, and they need to undergo an authorisation process described in the EU regulation (Articles 15-18).

Health claims other than reduction of disease risk. These claims can refer to the role of a substance in growth, development and the functions of the body; psychological and behavioural functions (e.g. 'lowers the blood cholesterol level'), and slimming or weight control/reduction in the sense of, for example, increasing the sense of satiety. Some authors refer to these as *Enhanced function claims* (Lähteenmäki *et al.* 2007). Such products are often enriched with some nutrients like vitamins or minerals. These claims can be made without the authorization process, as long as they are included in the list provided by the Commission, and are backed up by scientific data and understood by the average consumer. This list will be finished and available from 2010, after consultations with EFSA.

4.1 Studies on health claims

Van Kleef *et al.* (2005) found in a study that health claim was the main driver of the intention to buy different functional foods. They also found that consumers prefer health claims that are related to disease reduction or health enhancement. Consumers also seem to be more influenced by risk-related messages than the health enhancement claims, but this relationship is likely to vary among claim content and the benefit. Another study found that consumers do not perceive a difference in the type of health claim as long as the component offering the benefit is familiar (Urala *et al.* 2003). Lähteenmäki *et al.* (2007) on the other hand, report that claims related to improved health condition received higher approval than risk-reducing claims in the United K.ingdom, but the degree of approval is product dependent. The evidence also points towards personal relevance in the form of illness as having an influence on the attractiveness of health claims.

Consumers seem to prefer short, common language risk reduction claims, not long and detailed declarations. Wansink *et al.* (2004) tested two claims on a soya burger: a long (25 g of soy protein a day, as a part of a diet low in saturated fat and cholesterol, may reduce the risk of heart disease) and a short (Soy protein may reduce the risk of heart disease). The short health claims were shown to be more efficient and have a more positive effect on product evaluation than long health claims. The consumers seem to generate more positive inferences from short claims. Urala *et al.* (2003) found that consumers perceive short, simple messages ('contains...') as equally advantageous as longer, disease preventive messages for familiar substances (fibre, salt, etc.). There was, however, a difference for more unfamiliar substances like CLA, where more information seemed to enhance the perceived benefit from the product. Other research shows that consumers may have difficulties in understanding even simple nutritional messages that do not directly link the product to the health benefit (Lähteenmäki *et al.* 2007).

Trust in health claims is an important factor influencing functional food choice. Verbeke (2005) found that believing in the health effects of functional food products is the most important factor influencing acceptance. Consumers in some studies are found to be sceptical of health claims from manufacturing companies (Bhaskaran and Hardley 2002). Trusted sources are doctors, dieticians, educational institutions and family members, or the claim should be approved by the government (Williams 2005). Consumers in the US, on the other hand, seem to trust medical sources and media most: only 4% say that they trust scientists (IFIC 2005), while the Finns also trust the information on labels; they are also more positive about functional foods than most other studied countries (Urala 2005). The most trusted information sources in Finland were food and health authorities, while the least trusted were food manufacturers (Urala *et al.* 2003). But even so, the manufacturer claims do influence the purchasing decisions, and are used as an important source of knowledge.

Although some research indicates that the type of health claim may have an influence on consumer perceptions, Van Trijp and Van der Lans (2007) found that type of health claim does not influence consumer perception of overall healthiness of the products or specific health impact. The different claim types seem, however, to have an impact on how consumers perceive newness of a product, credibility of the claim and difficulty in understanding a claim, which again could have implications for nutritional education (Van Trijp and Van der Lans 2007). Content claims ('contains...'), which are nutritional claims, were found to be the most credible among the consumers, while the product claims ('helps bodily function') were least credible (Van Trijp and Van der Lans 2007). Product claims were also easier to understand compared to content claims. This depends, however, on the type of benefit. However, Van Trijp and Van der Lans (2007) conclude that the effects of claim types on consumer perceptions are not consistent across benefits and ingredients. There might also be differences between different carrier products. Thus, consumers are open to a wide range of health claims.

5. Carrier products

The carrier products have been shown to have influence on how functional foods are perceived by the consumer. There is also some evidence that the evaluation of health claims may be influenced by the carrier product (Bech-Larsen and Grunert 2003). The type of carrier product has also been shown to have the largest effect on consumers' perception of healthiness and willingness to buy functional food (Ares and Gambaro 2007). There are two main issues related to carrier products which are of importance: how healthy and natural the products are perceived to be and the potential effects of enrichment on sensory properties of the product.

There are, however, clear limitations in the EU regulation about which type of products can be marketed with health claims: they need to fulfil certain nutritional profile before claims can be attached to them: basically, unhealthy products cannot be carriers of health claims.

5.1 Perceived healthiness and naturalness

Perceived healthiness of functional foods is one of the most important motivations for choosing functional foods. Bech-Larsen and Grunert (2003) found that the type of health claim, processing method, enrichments and product types have a strong influence on the health perception of a product. Their findings suggest that basic products perceived as unhealthy (e.g. spread) would gain more from enrichment compared to healthy products. Unhealthy products cannot, however, be marketed with health claims according to the EU regulations. The authors also suggest that consumers will probably not accept functional enrichment of foods which are perceived as inherently healthy, which might pose a problem for marketers of marine functional foods. For example, enriched juice and yoghurt were perceived negatively when enriched with omega-3 fatty acids, while enriched spreads (which are perceived as inherently unhealthy) were perceived positively.

Others have found that consumers see products that are intrinsically healthy as the most credible carriers of functional messages (Poulsen 1999; van Kleef *et al.* 2005; Ares and Gambaro 2007). Consumers have more positive attitudes towards enriched product variants where the substances are already present in the conventional product (for example, bread enriched with fibre, enrichment of dairy products with calcium and vitamin D).

Healthy foods are often described as natural, pure, basic and unprocessed by consumers (Niva 2007). People seem to differentiate between healthy foods and functional foods: some people are afraid of getting an 'overdose' of vitamins if they eat enriched foods instead of, or in addition to, fruit and vegetables. Also, in Niva's (2007) study healthy food was never associated with hedonic pleasure: health and pleasure are often perceived as two opposites. The discovery that people prefer naturalness in food as opposed to enrichments is also confirmed by the findings of Rozin *et al.* (2004). They found that there are several types of beliefs related to the preference of naturalness. One is a moral belief: when people interfere, they cause damage to nature. A second belief is that natural food is healthier. The sensory properties are also perceived to be better in natural food. Finally, natural foods are considered as pure and safe. Devcich *et al.* (2007) also found that naturally healthy food is preferred to food with artificial 'additives'.

Foods implying the use of (new) technologies are often associated with negative perceptions like unnatural, distrust, unsafe, etc. (Bäckström *et al.* 2003). Functional food produced by using 'new' technology might cause consumers to perceive functional food as less natural than conventional or health products. There are also studies which indicate that consumers may have problems differentiating between functional foods and medicine (Urala 2005) while other studies have not found this relationship (De Jong *et al.* 2003).

5.1.1 Perceived healthiness of fish

Many studies show that fish is generally perceived as healthy (Olsen 2003; Verbeke and Vackier 2005). When consumers are asked in more detail about the nutrients or other beneficiary substances present in fish, they become uncertain. Verbeke *et al.* (2005) found in a study in Belgium that consumers generally believe that fish is nutritious, and that fish consumption reduces the risk of coronary heart disease. The consumers did not, however, know that fish consumption may stimulate brain development or that fish contains omega-3. The consumers also believed that fish contains a lot of dietary fibre, while the knowledge of the positive effect of omega 3 fatty acids on health was generally low. So there is evidence that at least for fish, the factual knowledge about specific nutrients and their effect on health is rather poor among consumers. The perceptions and knowledge vary for gender, age, whether people have children, and education.

All in all there also seem to be conflicting findings concerning the importance of the carrier product. This would seem to necessitate studies at the product level, not necessarily at the marine functional foods level.

5.2 Hedonic aspects

Few consumers are willing to compromise taste for health in food (Tepper and Trail 1998; Frewer *et al.* 2003; Verbeke, 2006). It can be a real risk if food enrichment influences the sensory characteristics of food (Tuorila and Cardello 2002), or is at least perceived to do so by the consumers (Verbeke 2006). Liking and taste of a product have been found to guide food choice strongly (Tuorila and Cardello 2002; Urala 2005). People may be sceptical especially towards seafood-based enrichments in other foods because of the expectations of a strong seafood odour or taste (Hillam 1998). Also, if some substances like fibre are added to fish, it may influence the texture, which can again reduce consumer acceptance of such products.

5.3 Price

Consumers in many countries are extremely sensitive to changes in food prices. Functional food tends to be more expensive compared to similar conventional products. Some studies show that there are consumers who are willing to pay more for functional products (Poulsen 1999) while others claim that price could be a potential barrier to the purchase of such products (Lyly *et al.* 2007). The issue of price can be especially interesting for fish as functional food: in many countries fish is considered to be an expensive food, and is already a barrier to fish consumption (Verbeke and Vackier 2005). An even higher price may create difficulties in launching marine functional products. This is an issue which should be studied in the future, especially since it is likely that there are consumers who are willing to pay for health benefits.

6. Demographic and cross-cultural differences in attitudes/ acceptance of functional foods

Gender, age and geography seem to influence how willing consumers are to accept functional foods (Bhaskaran and Hardley 2002; Urala and Lähteenmäki, 2004; Bech-Larsen and Scholderer 2007).Women are reported to be more health-conscious than men (Childs and Poryzees 1997; Poulsen 1999; Beardsworth *et al.* 2002), and are also a target group for functional food marketers. It seems, however, that this effect can be due to product-related differences between genders (Urala 2005). Women seem to perceive health-related claims more positively than men (Urala *et al.* 2003). Another study (Urala and Lähteenmäki 2007) did not find any differences in attitudes between men and women. Bech-Larsen and Scholderer (2007) suggest that the reason why women are more health oriented than men is that they feel more responsible for the well-being of their family.

Older people tend to find functional foods more rewarding than young people (Urala and Lähteenmäki 2007). They are also reported to be more interested in and more willing to accept functional foods (Tuorila *et al.* 1998; Verbeke 2005) than others. Older consumers tend to think more in preventive terms (reduce risk) while only some of the younger people say that health concern influences their decisions: their concern is then mostly related to weight (lean products, low fat, light, low in sugar) (Bhaskaran and Hardley 2002). Older people are also more health oriented because they might have a diagnosis of a life-style related illness or they know someone who has such illness, thus making health a very relevant issue (Bech-Larsen and Scholderer 2007).

There are also differences in the acceptance of functional foods between countries (Frewer *et al.* 2003). This may partly be a result of previously different practices and regulations related to health claims in different countries. Consumers in countries with previous exposure to health claims, tend to be more willing to accept functional foods because of their familiarity. Finland and Sweden are examples of countries where functional foods have been on the market for some years now, and the acceptance of new functional foods tends to be higher. Health claims have not been allowed in Denmark before the EU legislation. In the United K.ingdom, regulations have been stricter than those in Finland (no references to disease have been allowed).

In the US and Great Britain, people prefer to enhance their health by eating more fruit, vegetables, and fibre rather than taking supplements (IFIC 2005). On the other hand, they prefer supplements to functional foods (Frewer *et al.* 2003). The Danish are sceptical towards functional foods in general. Poulsen (1999) found that Danish consumers tend to perceive functional enrichments as 'artificial additives' which has a very negative connotation in many other countries as well (for example, Russia). The Finns have quite a positive attitude towards functional foods (Bech-Larsen and Grunert 2003). They differ from many other European consumers also in that they trust most sources of health claims (also food manufacturers), and they do not differentiate between whether the claims are risk reducing or health enhancing (Urala 2005). Belgian consumers are also quite critical about the concept of functional foods (Verbeke 2006).

7. Concluding remarks and need for further research

Studies on consumer acceptance of functional foods show some conflicting results regarding several of the important aspects. The significance and role of knowledge is important for consumer acceptance of functional foods. Knowledge level is related to the issue of which type of health claims should be used in marketing marine functional food: consumers have a different ability to process information depending on, among other factors, their knowledge level of the relationships between nutrients, food and health.

There seems to be no doubt that consumer attitudes towards different functional food products can vary; the weight of relevance, trustworthiness and safety can vary, making it necessary to study acceptance at a product level. Consumers do not seem to regard functional foods as a product class, rather they are evaluated together with other products in the same product category (e.g. functional yoghurt with other yoghurts). In all circumstances, the food should have excellent sensory properties in order to be accepted: very few consumers are willing to compromise taste for healthiness in food.

There has been some conflicting media coverage on the benefits and risks of seafood and seafood-based supplements which may add to consumer confusion and lower the acceptance of marine functional food. Warnings of high levels of PCB and dioxins in farmed salmon on one hand and the recommendation to consume seafood twice a week on the other hand creates uncertainty among the consumers: who can eat which type of seafood and how often? What kind of trade-offs are consumers willing to make, if any? Another discussion has been on the effect of omega-3 taken as a supplement or by eating fish: which is best, and are the supplements beneficial at all? Whose information can be trusted? The uncertainty among the consumers, and the level and influence of this confusion is likely to vary among countries. In countries like Norway, where consumers trust the health authority information on food safety, the effect is probably not large. In other countries like Germany, the trust in governmental information is quite low, and the effect of conflicting information is likely to be higher.

To sum up, there are some possible special concerns with *marine* functional food and consumer acceptance that needs to be studied further:

a. Seafood is perceived as a healthy food in itself by most consumers. Earlier research indicates that enriching inherently healthy products (such as seafood) might not be accepted by consumers. Also, enriching seafood might create a problem with naturalness. On the other hand, inherently healthy products are perceived as the most credible carriers of health claims. This area needs more research: how does the healthy image of seafood influence consumers' perceptions and acceptance of different enriched seafood products? How does a general health involvement influence acceptance of marine functional food? Also, how important is health as a motive compared to other motives in functional food choice?

Even though seafood is considered as healthy food, it faces some barriers among consumers: smell, availability of fresh fish, bones, high price, and family preferences. One important future research theme is therefore how a potential high price (assuming that enriching seafood with functional ingredients would increase the costs) would influence consumer acceptance for marine functional food?

b. Marine ingredients as components in other foods could be problematic because some consumers will think it unnatural to mix different product groups, and some might expect a negative influence on taste. This calls for further research: what are the perceived effects of marine ingredients in other foods? Which combinations will be accepted by consumers?
c. Consumer knowledge of marine functional ingredients and their health effects is quite poor. Yet, it is crucial for the consumer understanding of the benefits of functional food/ingredients. There is a need for further research on both subjective and objective knowledge about marine functional food/marine ingredients among consumers.

   We also need more research on the consumers' awareness of the relationship between the specific healthy components in seafood and their effects. Earlier research shows that seafood is perceived as healthy food, but that people don't know exactly why it is healthy. The role of conflicting research results on the healthiness of seafood and ditto media coverage on consumer perception is also an interesting future research area.
d. Research on consumer attitudes towards *marine* functional food is almost non-existent at the moment. We need more research on the attitude dimensions, which were reported in Finnish studies (Urala 2005), related to specific marine functional products. Studies about attitude formation are also of interest: what is, for example, the role of risk/benefit assessment in consumer attitude formation and purchasing decision? Are marine functional foods perceived as trustworthy by the consumers?
e. The interplay of attitudes, knowledge, carrier products and health claims related to marine functional food should be studied further. Based on this research, effective communication strategies for introducing marine functional foods to consumers can be designed.

References

Ares G, Gambaro A. 2007. Influence of gender, age and motives underlying food choice on perceived healthiness and willingness to try functional foods. Appetite 49(1):148-158.

Beardsworth A, Bryan A, Keil T, Goode J, Haslam C, Lancashire E. 2002. Women, men and food: the significance of gender for nutritional attitudes and choices. British Food Journal 104(7):470-491.

Bech-Larsen T, Grunert KG. 2003. The perceived healthiness of functional foods. A conjoint study of Danish, Finnish and American consumers' perception of functional foods. Appetite 40:9-14.

Bech-Larsen T, Scholderer J. 2007. Functional foods in Europe: Consumer research, market experiences and regulatory aspects. Trends in Food Science and Technology 18(4):231-234.

Bhaskaran S, Hardley F. 2002. Buyer beliefs, attitudes and behavior: foods with therapeutic claims. The Journal of Consumer Marketing 19(7):591-606.

Bäckström A, Pirttilä-Backman A-M, Tuorila H. 2003. Dimensions of novelty: a social representation approach to new foods. Appetite 40:299-307.

Childs, NM, Poryzees G. 1997. Foods that help prevent disease: consumer attitudes and public policy implications. Journal of Consumer Marketing 14(6):433-447.

De Jong N, Ocke MC, Branderhorst HAC, Friele R. 2003. Demographic and lifestyle characteristics of functional food consumers and dietary supplement users. British Journal of Nutrition 89:273-281.

Devcich DA, Pedersen IK, Petrie KJ. 2007. You eat what you are: Modern health worries and the acceptance of natural and synthetic additives in functional foods. Appetite, 48(3):333-37.

Frewer L, Scholderer J, Lambert N. 2003. Consumer acceptance of functional foods: issues for the future. British Food Journal 105(10):714-731.

Hillam M. 1998. The market for functional foods. International Dairy Journal 8:349-353.

Honkanen P, Olsen SO, Verplanken B. 2005. Intention to consume seafood - the importance of habit. Appetite 95(2):161-68.

IFIC. 1999. Functional foods: Attitudinal research (1996-1999). Washington: IFIC International Food Information Council Foundation. http://www.ific.org/foodinsight/1999/so/funcfdsfi599.cfm

IFIC. 2005. Consumer attitudes towards functional foods/foods for health. IFIC International Food Information Council. http://www.ific.org/research/upload/2005funcfoodsresearch.pdf

Lappalainen R, Kearney J, Gibney M. 1998. A pan European survey of consumer attitudes to food, nutrition and health: an overview. Food Quality and Preference 9(6):467-478.

Lyly M, Roininen K, Honkapää K, Poutanen K, Lähteenmäki L. 2007. Factors influencing consumers' willingness to use beverages and ready-to-eat frozen soups containing oat β-glucan in Finland, France and Sweden. Food Quality and Preference 18(2):242-255.

Lähteenmäki L, Lyly M, Urala N. 2007. Consumer attitudes towards functional foods. In: Frewer L, van Trijp H (eds.) Understanding consumers of food products. Cambridge: Woodhead Publishing Ltd. P. 412-427.

Niva M. 2007. All foods affect health: Understanding of functional foods and healthy eating among health oriented Finns. Appetite 48(3):384-393.

Olsen SO, Scholderer J, Brunsø K, Verbeke W. 2007. Exploring the relationship between conveninece and fish consumption: A cross-cultural study. Appetite 49(1):84-91.

Olsen SO. 2003. Understanding the relationship between age and seafood consumption: the mediating role of attitudes, health involvement and convenience. Food Quality and Preference 14:199-209.

Poulsen JB. 1999. Danish consumers' attitudes towards functional foods. Working paper No. 62. Aarhus: Aarhus School of Business.

Rozin P, Spranca M, Krieger Z, Neuhaus R, Surillo D, Swerdlin A. Wood K. 2004. Preference for Natural: instrumental and ideational/moral motivations and the contrast between foods and medicines. Appetite 43:147-154.

Steptoe A, Pollard T. Wardle J. 1995. Development of a measure of the motives underlying the selection of food: the food choice questionnaire. Appetite 25:267-284.

Tepper BJ, Trail AC. 1998. Taste or health: a study on consumer acceptance of corn chips. Food Quality and Preference 9:267-272.

Tuorila H, Andersson Å, Martikainen A. Salovaara H. 1998. Effect of formula, information, and consumer characteristics on the acceptance of a new snack food. Food Quality and Preference 9:313-320.

Tuorila H, Cardello AV. 2002. Consumer responses to an off-flavor in juice in the presence of specific health claims. Food Quality and Preference 13:561-569.

Urala N. 2005. Functional foods in Finland. Faculty of Agriculture and Forestry/VTT. VTT publications 581. Helsinki: University of Helsinki.

Urala N, Arvola A, Lähteenmäki L. 2003. Strength of health-related claims and their perceived advantage. International Journal of Food Science and Technology 38:815-826.

Urala N, Lähteenmäki L. 2003. Reasons behind consumers' functional food choices. Nutrition & Food Science, 33:148-158.

Urala N, Lähteenmäki L. 2004. Attitudes behind consumers' willingness to use functional foods. Food Quality and Preference 15:793-803.

Urala N, Lähteenmäki L. 2007. Consumers' changing attitudes towards functional foods. Food Quality and Preference 18:1-12.

Van Kleef E, Van Trijp HCM, Luning P. 2005. Functional foods: health claim - food product compatibility and the impact of health claim framing on consumer evaluation. Appetite 44:299-308.

Van Trijp HCM, Van der Lans IA. 2007. Consumer perceptions of nutrition and health claims. Appetite 48(3):305-324.

Verbeke W. 2005. Consumer acceptance of functional foods: socio-demographic, cognitive and attitudinal determinants. Food Quality and Preference 16: 45-57.

Verbeke W. 2006. Functional foods: consumer willingness to compromise on taste for health? Food Quality and Preference 17:126-131.

Verbeke W, Vackier I. 2005. Individual determinants of fish consumption: application of the theory of planned behavior. Appetite 44:67-82.

Verbeke W, Sioen I, Pieniak S, Van Camp J, De Henauw S. 2005. Consumer perception versus scientific evidence about health benefits and safety risks from fish consumption. Public Health Nutrition 8(4):422-29.

Wansink B, Sonka ST, Hasler CM. 2004. Front-label health claims: when less is more. Food Policy 29:659-667.

Williams P. 2006. Annotated bibliography of publications on consumer use of health claims. Report, National Centre of Excellence in Functional foods, http:www.nceff.com.au

Williams P. 2005. Communicating health benefits - do we need health claims. Australian Journal of Dairy Technology 60(2):192-194.

Commercialising healthy nutrition: challenges for SMEs

Patricia Wiklund
Invenire Market Intelligence Oy, Kokkilantie 835, FI-25230 Angelniemi, Finland

1. Introduction

Health is becoming a prerequisite for companies working with food and nutrition. All over the world, multinationals, as well as small and large companies are adjusting their strategies to take into consideration the growing interest among consumers to adapt healthier lifestyles. Health has become a fundamental concern affecting all food categories and strongly influencing the strategies of most food companies. Being aware of the growing problems of diabetes, obesity, heart and gut problems, and many other welfare diseases, food manufacturers are realising that it is time to adjust the sails.

In this chapter the role of small and medium-sized enterprises (SMEs) active in healthy nutrition within the Nordic countries is discussed. The characteristics, capabilities, challenges and opportunities of the SMEs in the marketplace are mapped. This chapter is partly based on a project within the Nordic Innovation Centre's programme Functional Foods, conducted during 2006-2007. Throughout the project a better and deeper understanding of the SMEs was gained regarding the issues of innovation acceleration, entrepreneurship, strategies to enter the market, and best practices. The implications of the general outcomes of this study for SMEs working with marine-based healthy products are also discussed.

2. The market of healthy nutrition

A term often used to describe health-beneficial products is 'functional foods'. In our opinion, the term is too narrow, interpreted in various ways, and not descriptive enough to capture the whole phenomenon of food and health. Therefore in our work we preferred to use the broader term, healthy nutrition. With healthy nutrition we mean any nutritional products and services that are focused on health, communicating health or supporting commercialisation of health. This may include functional foods, functional ingredients, health ingredients, technology, services, dietary supplements, health-enhancing foods, or, just intrinsically healthy foodstuffs or ingredients.

The market for healthy nutrition is best described as a large number of different niche markets with a common denominator in health. The products range from highly specific 'solutions' for targeted niche markets such as gluten-free products for celiac disease, to mass market solutions suitable for most consumers, e.g. vitamin D fortified dairy products and wholegrain.

In many markets for healthy nutrition, it is noticeable that SMEs have been quite successful. They are often highlighted as important players in the business, because they seem to have the vision, determination and endurance required to successfully bring new concepts to market. In general, however, little is known about SMEs active in healthy nutrition and health ingredients. Looking at the Nordic companies, there has been very limited information available on their commercial activities, market positions, products and services, successes and failures, innovation drivers and hurdles, etc.

Following the news about the withdrawal of healthy nutrition products, one has the impression that commercialising healthy nutrition is difficult. At the same time it represents one of the most exciting business areas for the food and beverage industry worldwide. Failure is commonplace and profits elude many companies. This can probably be explained by a number of reasons. In the Nordic countries there seems to be strong competence in applying science and technology, but commercial aspects of the innovation process are often neglected and promising inventions fail to succeed. This may be due to a lack of marketing insight and limited understanding of the business of healthy nutrition, as well as of markets and of consumer attitude and behaviour. Another reason for failure might be that inadequate resources are allocated for communicating, promoting and sustaining the invention along the different stages of commercialisation.

3. Purpose and methodology

The purpose of this study was to generate information and insight about the Nordic SMEs active in nutrition and health, their commercialisation activities, hurdles, drivers and needs. At the same time, the aim was to support policy-makers in setting up strategies to better support the SMEs to gain new insights about their business and how to strengthen their commercialisation processes and marketing capabilities. The project included a few major steps, which are depicted in Figure 1.

By means of desk research, monitoring and interviews country-specific studies were performed. An overview of activities and initiatives for supporting SMEs active in healthy nutrition was drawn up and companies working with healthy nutrition were identified and described. Based on this collection of company details, a survey of the Nordic companies was conducted. The main objective was to get insight into the commercialisation and marketing activities of the SMEs. The survey was carried out electronically with the use of survey software. This survey contained both multiple choice questions and more qualitative open questions. In total, 222

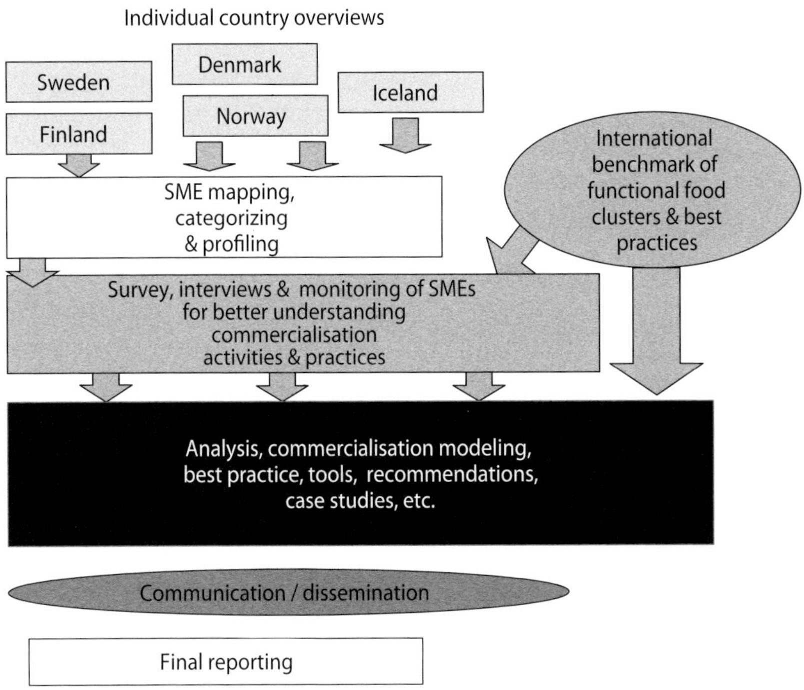

Figure 1. *Structure of the project.*

companies located in the Nordic countries were invited to participate in the study. A total of 50 questionnaires were returned, resulting in a response rate of 23%. Finnish and Swedish companies were the most active participants in the study and together they account for 76% of the study respondents. Five respondents were not involved in healthy nutrition and were excluded from the study, resulting in 45 valid replies.

Twenty Nordic SMEs were chosen for in-depth interviews. The companies illustrated different strategic approaches to bringing healthy nutrition to market. Most of the interviews were conducted at the companies' premises. They were semi-structured, audio and video recorded, and conducted with representatives at the highest management level of the company (mostly general managers).

A benchmark study was conducted to identify and analyse functional foods centres and initiatives internationally. This part of the study constituted both desk research and primary data collection by means of interviews with industry, academia and government officials.

Finally, all the data gathered during the project were analysed in order to present the information about the Nordic SMEs in terms of their commercialisation and innovation processes (hurdles, drivers, time span, strategies, products and services launched, level of internationalisation, support, etc.). Our general hypothesis was that Nordic SMEs lack the capability to fully manage, exploit and develop the innovations to reach their full commercial value.

4. Results

4.1 Who are the SMEs?

Small and medium-sized enterprises are defined by the European Commission (2008) as independent enterprises that have fewer than 250 employees and an annual turnover not exceeding 50 million EUR or a balance-sheet total not exceeding 43 million EUR. In addition, not more than 25% of the capital or voting rights should be held by another enterprise.

There is no distinct profile of a Nordic SME working with healthy nutrition. The companies vary in size, form, background, age, nature and goals. The diversity is striking. Most of them are micro companies, employing only a few people. Larger companies are mainly present in the area of food and beverages as well as dietary supplements, where production operations understandably require a larger workforce.

This study showed that there are few companies of the size of approx. 50-100 employees active in healthy nutrition. This is considered to be weak in terms of innovation and commercialization. In our opinion larger SMEs are important drivers for taking on new innovations and bringing them to market. Smaller companies often lack the resources, whereas bigger ones again tend to be more risk averse. These medium-sized enterprises would also be important customers for newly established ingredient companies trying to gain acceptance for their new solutions. The threshold for a 50-100 employee company to try something new from a lesser known supplier tends to be lower than that of a larger corporation.

In our opinion smaller enterprises should be encouraged to apply growth strategies and aim at becoming strong niche players within well-defined areas of healthy nutrition. This would increase the competitiveness of Nordic companies within the field.

4.2 SMEs' viewpoints on health

The Nordic SMEs are focusing on a large spectrum of health benefits. The ten most important areas are:

- cholesterol management;
- intestinal health;
- immunity;
- general heart health;
- blood sugar management;
- weight management;
- diabetes;
- children's nutrition;
- energy/vitality;
- sports nutrition.

An overall impression is that the Nordic SMEs have a very positive view of their business. Most of the survey respondents reported that their sales have increased during the past 12 months. In their opinion, their main products or services are very innovative in terms of health benefits. They also stated that the health benefits of the products were supported by solid science.

Nordic SMEs seem to be looking at health from quite a broad viewpoint. Products are often aimed at consumers, who are generally interested in health and wellbeing, although health is not their main buying criterion (Figure 2).

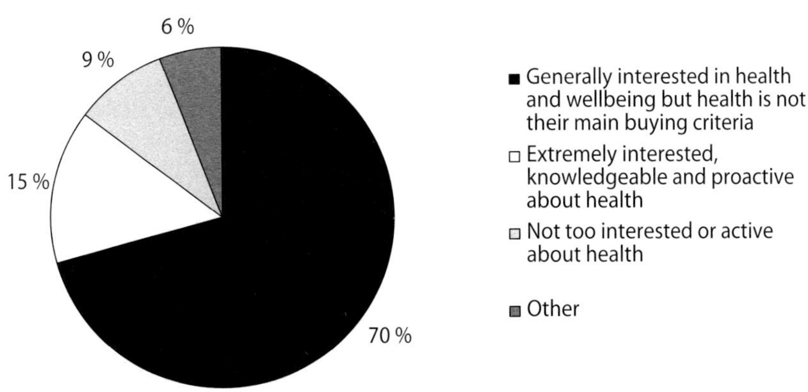

Figure 2. *Products are aimed at consumers, who are...*

4.3 Entrepreneurial attitude

In this project many SMEs of different sizes and forms have been involved. The wide spectrum of different approaches to commercialising has been striking. In-depth interviews were conducted with family-run established businesses with tradition and maturity, high risk taking ventures, university professors running small knowledge intensive companies as a side business, traditional food companies who have added value to their commodities through health, and ingredient and supplement firms who are in the very beginning of commercialisation. The background of the people in the SMEs were also very different; on a scale from employed professional managers to self-made business owners.

However, despite the great diversity, there seems to be a common factor in the most successful SMEs. The people in charge have a particular nature, which is probably best described as entrepreneurial. A definition that describes the SME managers in this project is given by Bridge *et al.* (2003):

> *An enterprising individual has a positive, flexible and adaptable disposition towards change, seeing it as normal, and as an opportunity rather than a problem. To see change in this way, an enterprising individual has security, born of self-confidence, and is at ease when dealing with insecurity, risks, difficulty and the unknown. An enterprising individual has the capacity to initiate creative ideas…develop them, and see them through into action in a determined manner. An enterprising individual is able, even anxious, to take responsibility and is an effective communicator, negotiator, influencer, planner and organiser. An enterprising individual is active, confident, purposeful, not passive, uncertain and dependent.*

This study showed that the adaptable disposition of the SME managers towards change and positive attitude was very dominant. Also, many SME managers were very good communicators, which seems to be an important tool in persuading and motivating others to share their visions and goals, and to get things done. On the other hand there is a risk of being overconfident, not adequately organised and, as a consequence, easily suffer from being under-resourced. An entrepreneurial manager is often in the risk zone of taking on too much and being too optimistic about his organisation's abilities. Another challenging issue is the question of power, and whether the manager is able to delegate tasks to others, or whether his urge to remain in control hampers the decision-making.

5. Product development race

The Nordic SMEs are actively launching new products on the market. Nearly all respondents said they had launched new products or services during the past three years. In addition, a majority was also planning new launches of health-enhancing products or services within the next twelve months:

- 88% of food & beverage and dietary supplement companies;
- 83% of the ingredient companies;
- 73% of the service companies.

The SMEs draw their inspiration for new product ideas from a variety of sources. As shown in Figure 3, most of them consider nutritional or technological research, market trends and market research as the most important sources for new product ideas.

The average time from the first phases of product development to product launch varies. The food, beverage and dietary supplement companies reported less than one year up to three years, while health ingredient companies also reported longer time spans up to more than four years.

In this study the question was raised as to whether the companies were putting too much effort into launching new products and services, particularly those working with consumer products. Maybe some of that effort would be better put into sustaining and improving existing products on the market. One manager expressed

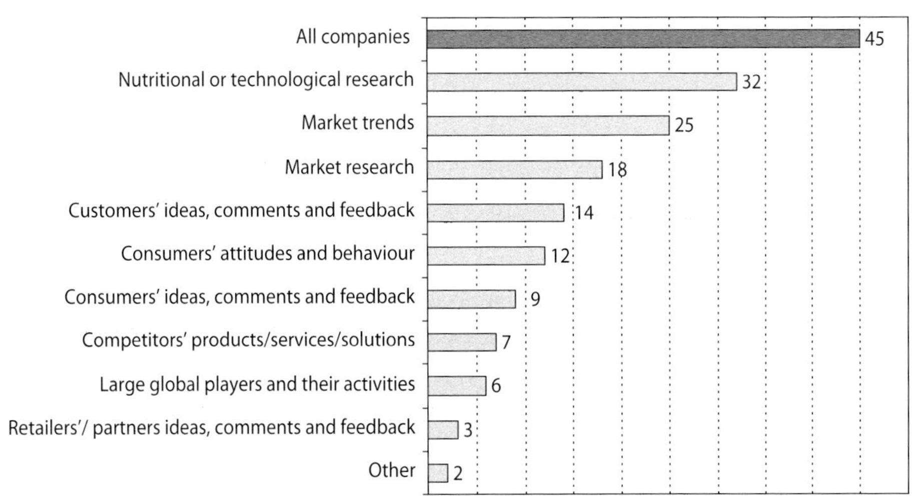

Figure 3. *Main sources for new ideas indicated by Nordic SMEs. The numbers indicate the times the factor has been mentioned. 45 companies answered the question.*

the fact that she experienced pressure particularly from the retailers to keep up the frenetic launch schedule. 'If you don't come up with something new, you are out.' Another manager commented on their past product development phases as too intensive: 'a company can only deal with a limited number of launches a year; if there are too many, the quality suffers and the organisation is exhausted'.

Other managers emphasised the importance of looking over the existing products and brands. 'We need to put more effort into positioning our products more clearly and brush up our company brand and what it stands for' was the comment of the marketing manager of an established dietary supplement company. Some of the SMEs had also developed alternative ways of dealing with the pressure of keeping up new product development.

A Swedish juice company had developed a new product together with the retailer in order to bring a new alternative to a promising product category. The process had entailed working out the concept, packaging and pricing together with the retailer. The manager described this kind of collaboration as good, even though he was critical about the unbalanced division of the risk. 'In these cases the retailer needs to take its own share of responsibility and not place all the risks on the SME developing the product. It is unsustainable for a retailer to place all the R&D cost on the company, without making any commitment to volumes or shelf space.' In this particular case the collaboration turned out well because the product seems to be well positioned, is strongly supported by the retailer's marketing and sells satisfactorily.

Another model identified in many newly established SMEs is the so-called spider-in-the-web model. The company outsources most of its activities and only holds the core capabilities and assets in house, e.g. the brand, certain patents or distribution. This strategy is applied in order to be fast moving on the market, and take less risk by not investing in own production and machinery, etc. The interesting new aspect here is the approach to product launches. This study showed one example of a Swedish dietary supplement company which developed the product concept by designing the recipe, the packaging, the layout and the communication. At this point the concept was introduced to potential clients. If the client showed interest then the product was manufactured and the contractors and suppliers were put to work. Controlling such a process based on outsourcing is challenging and many things can go wrong, as many players are involved. However, it seems to be an applicable model for small start-ups who are testing their concepts in the market place.

6. Commercialization drivers and hurdles

A commercialisation process can be examined by identifying a number of factors and events that either work in favour of or against the desired goal. During this project, the focus has been on identifying commercialisation drivers and hurdles, as they are decisive factors for the outcomes of the activities involved in bringing something new to market. The drivers and hurdles were explored both in the survey and the interviews conducted with SME managers. In the survey, the majority of the companies (69%) reported that the main drivers for commercialisation were consumers, market trends, scientific research, and resources such as staff, managers and capital. Whereas only a few companies mentioned that moves by competitors, ethical issues and public services were driving factors for commercialisation.

As can be seen in Figure 4, the major drivers mentioned are external. The first internal driver listed is competent and innovative staff. Obstacles and challenges for the commercialisation process of SMEs are mainly lack of funds and issues related to legislation and regulations (Figure 5). Food and beverage companies also found getting the taste and texture right, and negotiations with retailers and other

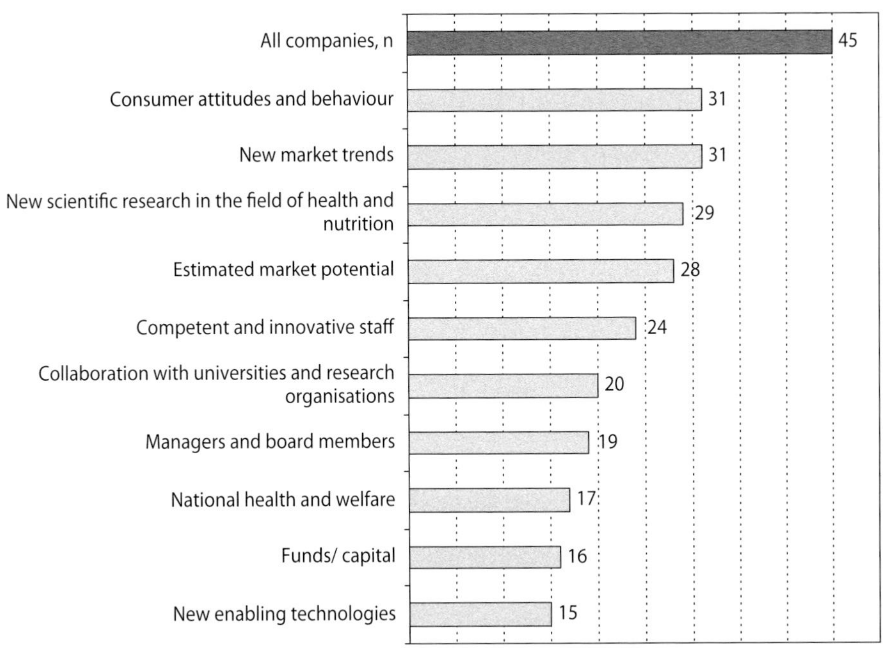

Figure 4. *Ten main drivers for the commercialization process indicated by Nordic SME companies within healthy nutrition. The numbers indicate the times the factor has been mentioned. 45 companies answered the question.*

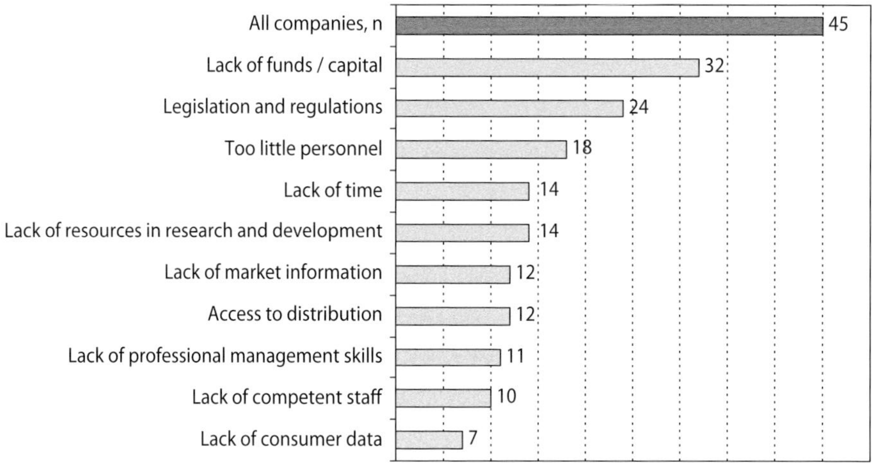

Figure 5. *Ten biggest obstacles in the commercialization process for SMEs indicated by Nordic companies. The numbers indicate the times the factor has been mentioned. 45 companies answered the question.*

distributors to be among the most challenging issues in bringing a new product to market. Ingredient companies replied that regulation issues, health claims, and communicating the benefits of the product were the most challenging issues in bringing new products to market.

It is a well known fact that almost 80% of all new food and beverage products launched fail. Healthy nutrition does not seem to be an exception. The requirements for successful commercialisation by the SMEs and their experiences concerning product failures were investigated in this survey.

This survey showed that over 40% of the food, beverage and dietary supplement companies have had to withdraw a health-enhancing product from the market over the past five years. In most of the cases, the withdrawal was made within two years after the launch. The most common reasons for withdrawals were unsatisfactory sales and lack of consumer understanding and appreciation for the benefits or taste of the product.

The situation for the ingredient companies is somewhat different. Only a few reported that they had withdrawn their product. The explanation can, however, be found in the different nature of the business. A poor selling ingredient can be kept in the product range for a long time even though it is not successful. Most common reasons for withdrawals of ingredients were unsatisfactory sales for the company or the distributor, and lack of consumer understanding of the benefits of the product.

A common observation is that many of the products fail because they are poorly positioned on the market and they fail to differentiate themselves from the masses. Many food and beverage producing SMEs have understood the crucial importance of taste. However, their marketing could be better planned and better resourced. Many of the interviewed companies reported the difficulty of understanding consumer behaviour and attitudes, as well as knowing which target group to focus on.

7. Marketing, branding and communication

Estimating skills in marketing, branding and communication is difficult as the measures are not easily defined, nor the results. When the companies were asked to evaluate their marketing skills, 47% argued that their skills in marketing were at a moderate level (Figure 6).

A majority of the companies would like to get better in building a brand. Other areas where respondents said they would like to improve were:

- targeting the right consumer or customer;
- identifying emerging new trends on the market or identifying customer needs;
- developing marketing communications.

Food, beverage and dietary supplement companies evaluated their skills to be best in managing the product launch, understanding and analysing consumer attitudes and behaviour and building a brand. Improvements were needed in building a brand, identifying emerging new trends on the market, targeting the right customer and marketing communications (Figure 7).

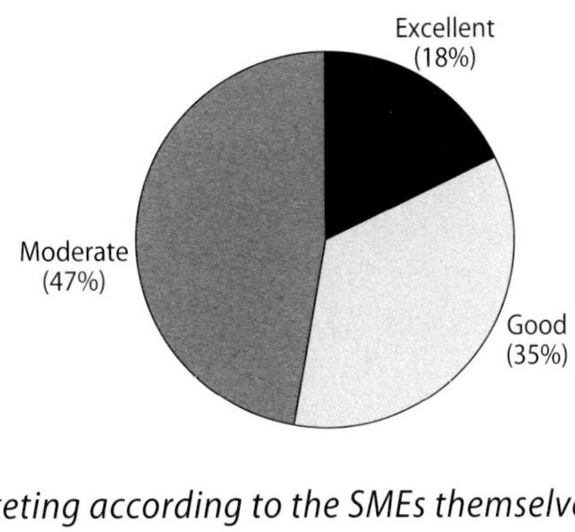

Figure 6. *Skills in marketing according to the SMEs themselves.*

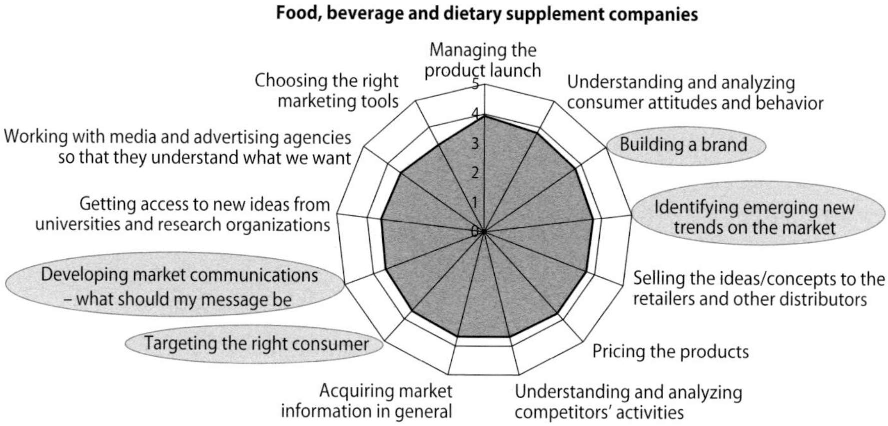

Figure 7. *Skills in marketing indicated by food, beverage and dietary supplement companies on a scale 5 excellent … 1 very poor. Circles indicate the main things companies would like to get better at.*

8. Discussion

Existing information about Nordic SMEs involved in healthy nutrition was minimal. There were no complete overviews in any of the Nordic countries. Thus, the initial task of gathering the basic information turned out to be a major one, and crucial for the continuation of the project. In other words, our initial premise that little is known about the Nordic SMEs proved right.

The SMEs involved in healthy nutrition vary in size, form, background, age, nature and goals. The spectrum was more varied than first anticipated.

The commercialisation and marketing competence in Nordic SMEs is quite low. This applies particularly to health ingredient and technology-based companies. Many lack a broad understanding of the market and have a limited understanding of consumer behaviour and the role of marketing and branding in creating value for the firm.

Nordic SMEs maintain a high rate of product development despite their small size and limited resources. To a certain extent this rat-race approach is exhausting the companies. It can best be described as a state of being where the companies feel continuous pressure to come up with new concepts and products. As a result they are doing a poor job in sustaining and supporting their current products.

Small enterprises in particular need to learn more about the management of the commercialisation process and what it entails. A main conclusion from this project is that many SMEs struggle to manage the commercialisation process professionally and keep their commercialisation activities in balance. As a result, many experience a constant lack of resources, stopping them from meeting their targets.

9. Implications for SMEs commercialising marine-based healthy products

Both research organisations and the industry claim that nutrients and other bioactive substances isolated from fish as well as fish in itself, have great potential in the area of healthy nutrition. From an SME viewpoint it is important to consider the key points in the commercialisation of marine-based products. The awareness of marine-based products seems to be rather low among consumers. Today many of them are no more than an ultra niche product. In order to successfully commercialise marine-based products SMEs must first promote their products and their health benefits. Preferably, with the help of the scientific community, government and health professionals, among others, in order to raise the awareness and knowledge about their benefits. As shown in this project, SMEs have limited resources and skills in marketing. Therefore, any efforts to support them in this task should be explored, such as the opportunities and potential benefits of joint marketing, PR and branding efforts. In the case of omega-3, which is the furthest developed marine ingredient in relation to health, there is a clear dietary need for omega-3, but it is still an ingredient that is quite unknown. Only in countries like Canada, the United K.ingdom and Norway, where it has received positive media coverage and well-oriented health benefit communication, has it succeeded in overcoming the niche status.

For the individual SMEs the challenge will be to develop a good and solid concept around its marine product and to address the following issues:

- **Convenience:** consumers are willing to pay a premium for convenience. Have we made our product easily accessible, easy to use and handle for consumers? Packaging is a decisive factor in this matter: a packaging that speaks convenience may be all that is needed to be successful.
- **Health benefit:** there is a growing body of research connected to the health benefits of marine products. The opportunity should be used to let the scientific community substantiate the health benefits. A challenge here is to identify which health benefit can be claimed. There may be many different options to choose from. In that respect one has to consider which health claim would be the most relevant for targeted consumers.
- **Marketing:** commercialisation will not work without a marketing strategy to enable credible communication to appropriately targeted consumer groups. The marketing efforts are decisive for success. A high level of consumer interest

generated by the media is probably the optimal situation for a company and its marketing.

Marine products have many of the major consumer trends working in their favour, of which the most important is, intrinsically, naturally healthy. Consumers follow many different routes to good health and they seem to prefer those products they experience as the most natural. The popularity of anything that is 'naturally healthy' is a marketing gem that SMEs should not overlook. Marine products have many health benefits built into them that can be commercialised, with or without heavy back-up from science.

10. Conclusions

Policy-makers wanting to help SMEs commercialise healthy nutrition need to gain better insight into the real hands-on problems that the companies are facing in their efforts. The tools offered need to be concrete, target-oriented and suitable for the SMEs. More action-based research would be recommendable to make the outcomes of scientific research easily available and comprehensible to SMEs. The commercialisation of science in connection with the situation of SMEs is an area that is recommended for further studies.

Nordic SMEs lack strength in brand building and developing good marketing concepts around their products. They do, however, have a good foundation in the technically strong products and well-documented research backup. Their marketing and commercialisation skills should be improved. The companies themselves need to employ brand developers and marketing experts with knowledge about health marketing in their organizations, and allow them to make a change. At the same time, support measures need to start focusing much more on the means to elevate the marketing and commercialisation capabilities and implementation activities in the SMEs. The support measures need to be well directed, hands-on and valuable to the SMEs. Due to their target-oriented and entrepreneurial approach, many small business managers need the guarantee of clear and concrete beneficial and commercial outcomes in return for their participation.

One of the main conclusions of this project is that even though there is much work needed to improve the commercialisation management and marketing competence of the SMEs, they are an important driving force for the industry. Due to the entrepreneurial character of many Nordic SMEs they fill the role of innovative and broad-minded players in the industry. With their boldness and tolerance of failure, they have successfully managed to bring new ideas to market that would have been dismissed as non-viable by more risk-averse and critically analytical larger corporations.

Nordic SMEs believe in their possibilities to be successful. They have an international reputation of being innovative within healthy nutrition. To keep this up the market needs further success stories.

References

Bridge S, O'Neill K, Cromie S. 2003. Understanding Enterprise, Entrepreneurship and Small Business. Palgrave Macmillan, United K.ingdom.

European Commission. 2008. http://ec.europa.eu/enterprise/enterprise_policy/sme_definition/index_en.htm.

Keyword index

A

B

C